Sound Affects

Also by Julian Treasure

How to Be Heard: Secrets for Powerful Speaking and Listening
Sound Business: How to Use Sound to Grow Profits and Brand Value

Sound Affects

HOW SOUND SHAPES OUR LIVES, OUR WELL-BEING, AND OUR PLANET

Julian Treasure

GRAND CENTRAL

New York   Boston

Copyright © 2025 by Julian Treasure

Cover design by Jason Booher
Cover image by Getty Images
Cover copyright © 2025 by Hachette Book Group, Inc.

Hachette Book Group supports the right to free expression and the value of copyright. The purpose of copyright is to encourage writers and artists to produce the creative works that enrich our culture.

The scanning, uploading, and distribution of this book without permission is a theft of the author's intellectual property. If you would like permission to use material from the book (other than for review purposes), please contact permissions@hbgusa.com. Thank you for your support of the author's rights.

Grand Central Publishing
Hachette Book Group
1290 Avenue of the Americas, New York, NY 10104
grandcentralpublishing.com
@grandcentralpub

Originally published in 2025 by Quercus, a Hachette UK company
First US edition: June 2025

Grand Central Publishing is a division of Hachette Book Group, Inc. The Grand Central Publishing name and logo is a registered trademark of Hachette Book Group, Inc.

The publisher is not responsible for websites (or their content) that are not owned by the publisher.

Grand Central Publishing books may be purchased in bulk for business, educational, or promotional use. For information, please contact your local bookseller or the Hachette Book Group Special Markets Department at special.markets@hbgusa.com.

Library of Congress Control Number: 2025934337

ISBNs: 9781538741870 (hardcover), 9781538741894 (ebook)

Printed in Canada

MRQ-T

Printing 1, 2025

To my beloved Jane, Holly and Sapphire,
who give me such loving listening.

CONTENTS

Listen!   1
A Sounding   3

Chapter One: How We Hear   15
Chapter Two: How Sound Affects You   27
Chapter Three: Sound Understanding   53
Chapter Four: Biophony   73
Chapter Five: Anthropophony   151
Chapter Six: The Sound of Tomorrow   205
Chapter Seven: Geophony   225
Chapter Eight: Cosmophony   257
Chapter Nine: Silence   301

Appendix   317
Scientific References   318
Further Investigations   319
Acknowledgments   327
Index   329

LISTEN!

I think I will do nothing now but listen,
To accrue what I hear into myself—to let sounds contribute toward me.

Walt Whitman, "Song of Myself"

This book is all about the urgent need for all of us to listen, partly for our own happiness, effectiveness and well-being, and partly because we are surrounded by extraordinary and wonderful sounds. To bring the auditory experience of the book alive, you'll find numbered references like this [123] throughout the book, each relating to a specific sound, from the songs of whales to the wind on Mars. The numbers correlate to links on the book's website, most of which allow you to enjoy listening to the sounds you are reading about. I hope they enchant you as much as they do me. A few of the numbers link to videos or reference sites.

All the books, films and music mentioned in the text are listed in my table of suggested further investigations, printed at the back of the book and also to be found on the book's website with links to help you find and explore them without doing the detective work.

Finally, for those of you with a scientific bent, the website also includes a list of the scientific papers and books underpinning many of the facts and findings reported throughout the book.

You can find the book's website at www.soundaffectsbook.com.

A SOUNDING

Deep beneath an unassuming squat concrete building in Redmond, Washington State, is the quietest place on Earth. Here, cocooned within a latticework of springs, steel and concrete—every wall a herringbone of fiberglass wedges—is the most efficient anechoic chamber ever constructed. Built in 2015 to test Microsoft's electronic devices, it holds the Guinness world record for the lowest ever measured noise at minus 20.3 decibels. This is as close to the absolute zero of sound as humanity has ever come.

I have spent time in anechoic chambers, and the experience is profoundly uncanny. With no external noise of any sort, the only sounds are those you produce. The thud of your heart, the tide of your blood moving around your body, your lungs expanding, each loud swallow, the workings of your digestive system, the soft creak of your joints... some have reported hearing the damp sound of their eyeballs moving in their sockets, while I have heard a hiss that's thought to be the spontaneous firing of auditory nerves. Most of these chambers set strict time limits for how long you can stay in them because the experience soon becomes disorientating, even frightening.

It is deeply unsettling to clap and experience the sound instantly disappearing. In a room without resonance, where we are not receiving the constant tiny auditory signals that inform us about our surroundings, even our balance is impacted. We begin to feel an odd disassociation from our surroundings. It doesn't feel like a room at all.

To return to a more familiar acoustic environment is to experience an overwhelming rush of auditory input and to realize anew how important and fundamental our relationship with sound is. It is this spirit that animates this book and that I hope to communicate to you.

*

We do not know exactly when consciousness arrives in a fetus, but we know its soundtrack—the steady (and surprisingly loud) pulse of our mother's heartbeat moving through amniotic fluid.* When each of us dies it will be marked by the silence following our heart's final beat. In between—from the almost imperceptible whine of a mosquito at the edge of our hearing to the beat of a nightclub felt deep in our chests, birdsong, waves on a beach, rain on a tent, the cowbell peal of a boat's rigging chiming across a bay, a baby's cry, a loved one's snore, the whine of a dentist's drill, or a jackhammer's staccato call—our lives are accompanied by a rich symphony of sound. For most of us, most of the time, it is background noise. We notice it when it becomes intrusive—a car alarm, a fox's screech in the early hours of the morning—or when it is a sound we enjoy and relish, perhaps the adrenalizing exuberance of Aretha Franklin, or the half-heard murmur of cricket commentary in an English summer garden. But, in between, there are vast amounts of sound we are

---

* Throughout this book I will often refer to the experience of sound and hearing as universal, though of course in reality the experiences vary enormously, especially for the hard of hearing. Though I write from within my own physical experiences of sound, I have no wish to downplay the validity of those very different relationships with sound in any way.

essentially unaware of. And they matter. Our acoustic environments impact our happiness, effectiveness and well-being. Who hasn't sighed in relief when an air-conditioning unit or the hum of a refrigerator ceases, even if it was never consciously noticed? That feeling of relief is a clue to the load that noise can place on us. There are things we can all be doing to improve our relationship with sound—but this is only part of the story.

Homo sapiens has been around for perhaps 300,000 years, and our ancestors for anything up to three million years before that. For the vast majority of that time, our relationship with sound has been first and foremost a matter of survival. Early humans used sound to detect and probably also trap the animals they hunted; even more crucially, like all living things, hearing was their primary warning sense. For most of human history, we've had good reason to fear the darkness. At night, at the edge of the firelight, our eyes begin to fail us—but our hearing functions just as well at night as in the day, and, what's more, we can hear what's behind us. As a result, sound goes very deep, very fast. If a twig snaps behind you in a forest you will spin around: the survival instinct is non-negotiable. If I want to warn you about urgent danger, I don't wave, I shout. This was brought home to me once in Kents Cavern, a cave system in Devon where they have found human remains dating back forty thousand years. In a small chamber deep underground, we sat down and switched off our torches. The darkness was thick to the point of suffocating, and my sense of hearing was suddenly dramatically heightened. It was easy to imagine sitting in that spot thousands of years earlier with only a small guttering torch, listening constantly for danger in the surrounding darkness. Since language developed at least 100,000 years ago, knowledge has been passed from generation to generation in sound, which is ephemeral and invisible: people learned by listening to teachers speaking, and if they missed it, they missed it. This is why Pythagoras is said to have erected a screen for his first-year students, who were called *akousmatikoi*, so that they could not see the teacher; he believed that seeing was a distraction from the important business of listening. For all but the last few thousand years of human history, our culture and communication were oral and aural.

In the fourth millennium BCE, writing developed independently in many places, although until the Industrial Revolution only the elite few in most societies were literate. In the short time since then, accounting for less than a tenth of one percent of human history, our relationship with the world has dramatically tilted and become heavily weighted to the eyes. Visual communication has become dominant. We teach reading and writing in school, but not speaking or listening. Of course, there are good reasons for the popularity of writing, which revolutionized how we stored and transmitted information, allowing us to send stable information through time and space in ways that changed humanity. Most of the communication methods we've invented since personal computers and the internet arrived have been visual: email, SMS, social media and instant messaging hold hostage our eyes and our fingers. Screens on mobile devices have further increased the dominance of the eyes as most people's primary portal to reality.

Much of our communication now is written without much regard to the consequences: quick, overheated opinions that will earn the tiny dopamine reward of a thumbs-up icon; typo-laden group messages; instantly regretted, half-thought-through emails, perhaps accompanied by the slightly surreal sound effect of an imagined physical message rushing through the air. For many people today, the experience of entertainment at home is a similarly distracted, multi-screen one, in which we're never entirely focused on one thing. We consume so much of our media in transit, whether literally or figuratively, which is why a recent study found that internet users are eighty percent more likely to watch a video if it has subtitles. The requirement to focus on audio has become almost countercultural. However, it is no coincidence that podcasts have seen their recent rush in popularity. This is an intimate, conversational medium that allows listeners to do something else while enjoying the show, something that's impossible with video- or text-based content. There are huge corporations spending billions on ever more seductive ways to seize our attention, almost always through the eyes. I met MIT professor Sherry Turkle, author of the excellent books *Alone Together* and *Reclaiming Conversation*, when she gave her important TED talk on the way technology is hollowing out our close relationships and replacing a few deep connections with many shallow ones, and we shared a deep concern about the future of intimacy and conversation. Nobody can restrain the unrelenting march of technology—almost nobody is even trying—but its effects on humanity's ability to listen have been, and continue to be, disastrous.

In our everyday lives, it is faster to read words than to speak them. Text-based communication is undeniably efficient because it's asynchronous—both ends don't have to be present at the same time. However, it often removes the human joy of the act of communicating and reduces it to an exchange of information. There is hope though: a reported recent increased preference among Generation Z for sending voice notes instead of text messaging could just herald a return to appreciating audio as a warm, authentic medium among those who have grown up with so much visual information.

Architecture, too, is almost exclusively ocular, resulting in many buildings that are so noisy they are unfit for the purposes for which they were built. Often, the harmful noise results simply from bad acoustics—typical of spaces that have been designed for the eyes only, with little or no consideration for the ears. Most of our cities are too loud, and many transport systems are literally deafening. If we were to accept the visual equivalent of our auditory experience it would be to travel with strobe lights permanently dazzling our eyes. It's no wonder that urban earbud-wearing "podestrians" try to take control of their auditory experience. Sound has become something to block out when we notice it at all.

Our world has become increasingly interventionist, a place where quietness, passivity or acceptance are often seen as weaknesses, and those who have the loudest voices or who deploy bullish power to affect things (for right or wrong) are preferred as influencers and leaders. Susan Cain's outstanding TED talk and her book *Quiet*—about the value of quiet, introverted people—give the lie to this obsession with intensity and loudness and propose that instead we should make use of the introvert's ability to think deeply, strategize and solve complex problems. She cautions: "Don't mistake assertiveness or eloquence for good ideas." In our busy competitive world, it's not surprising that urban consumer culture prefers seeing to hearing, sending to receiving, and speaking to listening. My own TED talk on speaking has at least five times as many views as my TED talk on listening—a ratio almost exactly replicated in research by the Organisational Listening Project, a research initiative that surveyed a range of organizations and found that they were devoting more than four times as many resources to outbound communication as they were to listening.

I believe that there is also a spiritual dimension underpinning this shift to the visual. Sight is an active sense: you decide where you are looking, you have eyelids you can close; and visual signals are analyzed in the cortex, the region of the brain we associate with higher-level processes. By contrast, hearing is passive: you have no earlids and you hear even while you sleep; as soundwaves enter the holes in your head, they physically touch your eardrums and activate the more primitive regions of your brain. Human culture is increasingly based on logic and agency—making things happen, as opposed to going with the flow. We've largely lost our connection with nature as we live in environments that are entirely constructed. And the same increasingly applies to our relationship with the invisible.

Archaeoacoustician Iegor Reznikoff reflects that for most of human history, life was short and tough, so people were more engaged with the invisible, especially with what happened after death—as a result, initially shamanism and later religion were central to their daily lives. Sound is invisible, and partly for that reason it played a key role in these practices—but we now live in a world where seeing is believing, and what's visible and tangible is preferred. We sanitize and even hide death; secularism and materialism encourage us to concentrate on the here and now, the object that can be seen and acquired. The visible increasingly dominates our consciousness and our choices. The eyes rule, while the ears are largely ignored.

I have spent the bulk of my professional life at the junction of sound and communication. I believe that reassessing our relationship with the sound we make and the sound we experience will have a profound impact on how we interact with one another and with nature. I define listening as *making meaning from sound*. This involves two stages: selecting what to pay attention to (which is largely done subconsciously by the brain's reticular formation), and then ascribing meaning to it. My belief is that increasing our awareness of sound and employing conscious listening can transform our happiness, effectiveness, well-being—and, in particular, our personal and societal relationships. Never have we needed listening more than we do today in a world threatened by polarization, dogma and conflict.

Hearing is a capability, but listening is a skill. It's tragic that we don't teach it in schools or places of higher education. Conscious listening can be learned, practiced and mastered, and the advantages for anyone who takes the time to do so are enormous. Listening is how we learn, and is the foundation of leadership, persuasion, sales, relationships and even civil society. It is increasingly true that we have stopped listening to each other and that internet-driven polarization and intolerance are becoming the norm. It's not a coincidence that we call online groups echo chambers, because they are spaces where people reinforce one another's prejudices, opinions are widely confused with facts, and conspiracy theories, post-truth politics and deep fakes distort and hide reality. In all of this, listening is ditched in favor of shouting down the opposition.

Attentive listening is a practice that enhances consciousness and forges fundamental connections in three important dimensions.

First, in time. Listening is often the main way we perceive the passage of time. Sound always exists in time: there is no such thing as an auditory photograph. The sound of music or speech acts like a ticking clock, and iconic sounds like the dawn chorus or a kettle boiling can mark important times of day. In his book *The Third Ear*, Joachim-Ernst Berendt quotes Hermann Hesse: "Music is time made aesthetically perceptible."

Second, in place, both physical and spiritual. Your ears are constantly giving you information about your surroundings: even with your eyes closed, you can sense the kind of room you're in and who or what is there with you, from tiny sounds and acoustic reflections. And spiritually, every tradition that I am aware of has at its heart quiet contemplation of some kind, involving silent listening to the person's connection with the ineffable, or to their own innermost essence, as appropriate.

Third, in society. All the research about happiness I've seen suggests that the foundations for personal happiness are a strong network of family and friends, and doing service for others—both built on relationships. All our relationships are in turn built on listening, which is why the most common complaint in any relationship is: "You never listen to me."

This is even more true of groups and organizations. Every human being listens uniquely, because we all listen through filters—the cultures we are born into, the languages we learn to speak, the values, attitudes and beliefs we pick up or set aside throughout our lives—and our listening changes over time and contexts. At any moment we may have expectations, intentions, emotions and assumptions that color our listening—for example, people tend to vote for politicians with deeper voices because we have the innate assumption that more significant (or possibly dangerous) things make deeper sounds. It is a grave but almost universal error to assume that everyone listens like you do. They don't. Simply understanding and accepting the diversity of other people's listening can transform communication and teamwork.

Sadly, this is rare in most organizations. I've already mentioned the Organisational Listening Project, a major piece of research carried out between 2014 and 2016 by a team headed by Professor Jim

Macnamara at the University of Technology, Sydney. They surveyed thirty-six organizations ranging from large to small across public, private and nonprofit sectors, and found that on average those organizations claimed to be devoting eighty percent of their time and resources to outbound communication such as PR and advertising, and only twenty percent to listening. As Macnamara noted: "Given that these are self-assessments, claims in relation to time and resources spent on listening are more likely to be generous rather than minimalist." We all like to think that we listen more and better than we really do. The conclusion of the research was damning: "Most organizations listen sporadically at best, often poorly, and sometimes not at all."

Of course, organizations don't have ears: the reason for this abject corporate listening is that they are full of people who are not listening. Macnamara found that there is solid financial justification for changing this. Paying proper attention to listening is a competitive advantage, because the survey found that organizations ranked as the better listeners had better staff and customer retention, higher morale and productivity, and better reputations. Those at the bottom of the listening table experienced more criticism and crises. Listening is profitable.

With this book I want to celebrate the miraculous story of sound, which takes us from the microscopic choir of the cells in our bodies to the largest cosmic scale we know of. The more we learn, the more we are discovering how alive with vibration everything around us is. The Perseus cluster is one of the largest structures in the known universe, containing thousands of galaxies surrounded by a superheated cloud of gas; at its center lies a super-massive black hole. In 2003 a team from Cambridge University analyzed fifty-three hours of readings from the Chandra X-ray Observatory space telescope and detected powerful soundwaves generated by this black hole, causing huge ripples in the super-heated plasma that surrounds it. They calculated the sound's frequency as once every 9.6 million years, which is a note some fifty-seven octaves below middle C. Though it is true that in space no *human* can hear you scream, an entity with an eardrum the size of a galaxy might be able to hear that note. Amazingly, as we'll explore later, even intergalactic space is full of sound.

It is a remarkable and thrilling story, and along the way I ask you to keep in mind that it's a story with a purpose—to help you reclaim sound as a vital part of your life, and by so doing make your life better. *Sound Affects* is about the effects of the sound we humans make, and the sound we, and all other living things, consume. Throughout the book, you'll find suggestions for practical actions you can take to improve your own happiness, effectiveness and well-being, and that of those you interact with—as well as actions we collectively can take with sound to support the increasingly fragile ecosystem of this planet that we so often take for granted. By becoming more conscious of sound, we can all take responsibility for how it affects us, and how we affect it.

Let's start by understanding what is happening when we hear.

1

HOW WE HEAR

If you want to find the secrets of the universe, think in terms of energy, frequency and vibration.

Nikola Tesla

For centuries, human beings have been invited to imagine a tree falling in a forest with no one around to hear it and to consider whether it makes a sound. For many of us, it's an amusing thought experiment we encounter as children, an introduction to the riddle of consciousness and perception. Throughout the twentieth century, as scientific theory and technology shifted, the pendulum swung on whether this is a simple question to answer or a complicated one. Currently, a perfectly reasonable answer would be: it depends.

However, I want you to imagine a different scenario: a tree falling in a forest when you *are* there to hear it. This can be whatever kind of forest you're most familiar with: a damp, deciduous European wood; a rich, warm equatorial jungle; a snow-ridged taiga; or anything in between. But as you walk slowly through it, you hear a mighty crack— and if you're sensible you'll stop walking and eye the trees around you with caution. But what is actually happening?

Formally, sound is vibration that propagates as an acoustic wave through a transmission medium such as a gas, liquid or solid. A simpler definition would be "audible vibration." A tree falls and as part of the release of kinetic energy, it vibrates, producing a soundwave that travels through the air then meets your ear, where in a physical and then electrochemical process, you perceive this as hearing a sound.

From a loved one's whisper to the eruption of a volcano, the fundamentals are the same. So let's unpack the miraculous process by which those audible vibrations become what we understand as sound.

*

To watch the percussionist Dame Evelyn Glennie play is a remarkable experience. She is arguably the first person to make a career as a solo classical percussionist. As her hands seem to blur on one of the many instruments she plays, finding and responding to complex rhythms, she seems entirely part of the music, a pure example of how organized sound connects us to something both deeply human and extraordinary. There is a fact that makes this even more remarkable: Evelyn Glennie has been profoundly deaf since she was twelve years old, after the nerves in her ears deteriorated. She has learned to sense sound and music by using different parts of her body.

This may seem counterintuitive. For most people, what happens when we perceive sound is intimately bound up with those seashells of cartilage, skin and fat we call our ears,* but we actually hear with our whole bodies. Our ears have evolved to hear better than any other part of us, but our bones, tissue, organs, and even our eyes, can hear.

Most people dislike hearing a recording of their own voice, which

---

* What we commonly call our ears are in fact only the first part of the ear and should correctly be called auricles, or pinnae.

often comes with the response: "Do I really sound like that?" That's because you are used to hearing your voice originate in your throat, and then resonate through your chest and head cavities; you hear it not only through the air and into your ear, but also via bone conduction in your skull. To hear your voice divorced of this other acoustic information transmitted through bone and tissue conduction is akin to hearing someone else entirely.

Hearing is the first sense we develop, as early as twelve weeks after conception—long before the ears are fully formed. What we hear in the womb is mainly our mother's heartbeat. Its three-time lub-dub-pause, lub-dub-pause is the rhythm of waltzes and lullabies. The tones and cadences of a mother's voice, muffled and bass-heavy in the womb, mean that when a child is born they can already distinguish their mother's voice (and language) from others. The shock of birth is at least in part the sudden revelation of sharp, high-frequency sound traveling through air—a dramatic change from the damped sound traveling through fluid that we have experienced until that moment—combined with the sudden removal of that comforting heartbeat. No wonder newborns cry.

Though it takes months for babies to learn to see the world, their ears are already fully active well before birth, so our early lives are dominated by the speaking and singing of our mothers. Mothers of every culture on the planet sing to their babies. Some anthropologists like Steven Mithen, a leader in the field of cognitive archeology, believe this powerful instinct may even be the very origin of language. Richard Parncutt, Professor of Systematic Musicology at Karl Franzens University in Graz, Austria, has proposed that in the womb we learn the links between our mother's patterns of sound and movement and the associated hormonal states we experience (because we share her blood flow) and suggests that these links are stored in "transnatal memory" and form the basis of our instinctive emotional responses to music. A baby's ability to take the sounds they are hearing and use them to form words of their own is a key early milestone. In sleep, a baby's eyes close but their ears remain open, as any parent of a newborn knows all too well.

Soundwaves are first gathered by the pinnae, which are optimized to amplify the frequencies of human speech. Vibrations are then channeled into a one-inch-long ear canal before reaching a delicate membrane the size of your little fingernail and a tenth of a millimeter thick, known as your eardrum. It is so responsive that the faintest sound we can hear moves the eardrum just two atomic diameters. That's almost sensitive enough to detect individual molecules of air hitting this extraordinary membrane.

Every sound you hear, from a symphony orchestra to raindrops on leaves, is then translated into minute pushes and pulls on three tiny, connected bones whose graphic popular names most schoolchildren remember: the hammer, anvil and stirrup. The hammer, or malleus, is pushed by the eardrum and transfers the vibrations to the anvil, or incus, which activates the stirrup or stapes, which in turn vibrates a membrane-covered opening known as the oval window. The three bones amplify the sound vibrations as they enter a tiny snail-shell-shaped structure full of liquid known as the cochlea. This contains an elastic membrane of collagen called the basilar membrane, upon which sit tiny hair-like structures whose tips have stereocilia, clumps of microscopic hair-like cells five times thinner than a human hair. As the vibrations ripple through the fluid, waves form along the basilar membrane and move these hair cells. As they rise and fall, channels at their tips open up, and chemicals rush into the gap, creating electrical signals. Hair cells toward the wide end of the snail-shaped cochlea detect higher-pitched sounds, while those closer to the center detect lower-pitched sounds. There are also three rows of outer hair cells, which protect and help sharpen the frequency response of the inner ones, effectively amplifying them. The movement needed to make a hair cell activate is tiny, roughly the diameter of a single hydrogen atom.

To understand what happens to the electrical signals once they are created, I talked to Jan Schnupp, Professor of Neuroscience at City University of Hong Kong and editor of the book *Auditory Neuroscience*. Jan explained to me that our neurons send impulses to one another through connecting axons at a typical speed of around thirty meters a second—a tenth of the speed of sound. This is painfully slow compared to the speed of electrical signals through copper wire, which move at close to the speed of light—though with neurons working in parallel, the brain is very power-efficient. Jan said: "It's slow compared to conventional electronics, but that's because we need to be able to power a supercomputer with a sandwich a day."

At each of typically five steps in the journey, there is a tiny synaptic delay as each neuron decides what to do with its input: every neuron connects to ten thousand others, so it has a lot of options. Nevertheless, the whole chain takes just two milliseconds to process high-frequency sound. This is over twenty times faster than our processing of visual input; because hearing is our primary warning sense, any sound needs to be assessed for potential danger as quickly as possible. It takes longer to perceive low frequencies, which is why whistling or clapping for urgent attention is a good idea.

Your auditory nerve fibers grow out of the cochlea and arrive at a part of the brain stem known as the cochlear nucleus, a kind of junction box, where they immediately branch out to different parts of the midbrain. Some fibers concern themselves with changes over time, which is how we perceive loudness, pitch and timbre; others process the difference between the signals from each ear, which is how we determine the direction of the sound source. The same signal is used for multiple analyzes, all at the same time.

Two important regions that receive auditory signals are the amygdala and the cortex via the thalamus. Your amygdala is often termed the limbic center, and is the part of the brain involved in immediate behavioral and emotional responses. It works fast, so certain sounds may provoke a physical emotional response well before your cortex has a chance to catch up. The sound of a car backfiring will have your heart racing before your cortex has worked out that there is no threat to your safety.

Brain imaging has also shown that some sounds—for example, scraping sounds like chalk on a blackboard, or metal on glass, which occur in certain high-frequency ranges—activate the amygdala as a kind of distress signal and are then perceived as unpleasant. There are suggestions that because such sounds are in the same range as human screaming or noises made by some predators, they may have been linked to pain or emergency for so long that they are now genetically associated with negative experiences.

It's important to realize that sound often affects you below the level of your consciousness. In environments with many different sounds, while your cortex is concentrating on a foreground sound there may well be something in the background that your amygdala is tuned into, causing you to feel stressed, agitated or nervous—another good reason for practicing the skill of conscious listening.

We will keep returning to this idea again and again, as it underpins so much of what we can practically do to improve our auditory environments. Sound is influencing you every day in ways your conscious brain isn't necessarily aware of, because of the way auditory signals are processed.

WHAT YOU CAN DO

I teach two exercises in my workshops that are designed to increase awareness of sound's impact. The first is called "the Mixer." Next time you're in a complex auditory environment—for example, a coffee shop, an office or a forest—pay attention and start to disentangle the individual sounds comprising the soundscape. Think of them like channels on a mixing desk in a studio, where each channel contains a separate instrument. You will start to hear sounds that perhaps were masked by others, or simply went unnoticed before. Making a habit of this practice, you will become faster and faster at discerning the components of your soundscape, and eventually move to a more conscious relationship with the sound around you.

The second exercise is called "Savoring." Just as you do with food and drink, set out to taste the sound around you. You would instantly spit out bad food, but we frequently endure unpleasant sound that's stressful or irritating, simply because we don't notice it. Savoring is testing sounds to see how good they are for you—a very personal practice, of course. You can take this into the spaces you normally occupy simply by closing your eyes and attentively listening to the sound, asking yourself: "Is this the best sound I could have in this space?" You may notice hums, buzzes or other noises that you have put up with for years; here is your chance to deal with them at last.

*

The elements of this complex process that we call hearing are all happening constantly, with tens of thousands of signals every fraction of a second, branching and registering in different places, changing as the sounds enter our ears. The difference between our boots breaking through the crisp top layer of snow and the softer crunch of the layer underneath; the sound of a baby's laughter shading into crying; a car's engine changing pitch as it passes us: we process and interpret all these different patterns of continuous signals and assign meaning to those we choose to attend to.

Our brains are optimized for spotting patterns, because any regular pattern tends to indicate agency by something living—and for millennia that meant either food for us, or us being food for something else. This innate desire to spot pattern and rhythm is probably one reason why music is so attractive to human beings. When the brain does perceive a pattern in sound, it will then try to predict what will happen next, as a self-preservation reflex. In the context of music, this means that our brains release dopamine when we hear music we like. We've all experienced the way a song that at first evinces little or no reaction becomes one we have to listen to again and again, as the promise and completion of the pattern become increasingly pleasurable once our brain recognizes it.

We process speech mainly in our left (rational) hemisphere and music in our right (intuitive) hemisphere. The brain stores every sound it hears as memory in several areas of the brain, so that we can recognize it in future and trim our response to be ever more effective. Hearing sounds that we have stored deep in our memory—for example, a nursery-clock chime from babyhood or a recording of a long-dead friend's voice—can evoke very powerful responses in us.

We would have trouble staying sane if all sounds were paid equal attention: life would be a constant cacophony and impossible to understand. Your brain does a lot of filtering so that you can understand the world around you. It's working all the time to differentiate signal from noise, so that you can understand the person talking to you in a noisy room full of other people talking. At the same time, a very old part of the brain called the reticular formation is at work discounting sounds that you have learned require no response, while bringing to your attention those that may. This is why you cease to be aware of air-conditioning noise after a short while but come to attention immediately if someone speaks your name.

The process of selection is highly associative: your life experience will guide your subconscious systems as to what to discard and what to pay attention to. Someone who has lived in the countryside will often be surprised by the constant city noises on a visit, while conversely city dwellers can find the quiet of the countryside disturbing. Some sounds demand attention from almost everyone—sudden noises, babies crying, alarm sounds and other people's conversation are the most distracting sounds of all—but the attention paid to many other sounds is entirely personal and can be adaptive. Hence some people can sleep soundly in nighttime traffic noise but will wake instantly if their baby cries, while soldiers can learn to sleep despite the sound of artillery.

In making such decisions, your brain will compare each sound to its database of sounds heard before, especially seeking any pattern it recognizes: your name is the sonic pattern that you learn to respond to earliest. Our auditory bandwidth is surprisingly limited, as you will know if you've ever tried to understand two people talking at once. However, musicians tend to develop a greater capacity for multi-track listening, because being a good musician means attentively listening to all the other players in a group or orchestra. Playing an instrument involves memory, translating visual cues (written music) into movement, adept motor skills and focused listening, which in combination use both sides of the brain intensively. As a result, it's now been established that musicians have larger brains than non-musicians, with more cells in the corpus callosum, which is the bridge between the left and right hemispheres, and in a region called the Heschl's gyrus within the auditory cortex, where professional musicians have 130 percent more neurons. In his fascinating book *Musicophilia*, Oliver Sacks wrote: "Anatomists would be hard put to identify the brain of a visual artist, a writer, or a mathematician—but they could recognize the brain of a professional musician without a moment's hesitation." Perhaps due to the extra neurons, tests also indicate that musicians have better memory—children with one to five years of musical training were able to remember twenty percent more vocabulary words read to them off a list than children without such training.

Hearing is unique among our senses in that it simultaneously perceives value (a note) and relation (an interval). We can't see when a color has a frequency exactly double that of another color, but we can hear that proportion as an octave. Asked to sing their favorite song unaccompanied, most people will sing in very good pitch, starting at roughly the right note: we can remember pitch as well as intervals.

WHAT YOU CAN DO

Playing music with others is a great way to enhance your listening skills as well as your social connections. It's never too late to take up an instrument and play with others, whether that's singing in a local choir, joining in a drum circle or playing in a band or orchestra. Encourage your children to learn an instrument and you are helping their brains to grow and improving their life skills.

All this neural activity happens in stereo: the brain is constantly processing two continuous signals, one from each ear, each slightly different due to the six-inch positional difference in location of the pinnae and the 180-degree difference in orientation. Animals with two ears locate the source of a sound largely by comparing the timing and volume of its arrival at each of the ears—which is, in fact, the whole point of having two. A sound from the side hits one ear directly but has to go around the head to arrive at the other; and the small difference in arrival time and volume is what our brain uses to calculate the direction from which the sound arose. In something called the superior olivary complex (SOC) the two signals combine, and our brain synthesizes them to create a three-dimensional soundscape. With our eyes closed, we can pinpoint the location of a buzzing fly as it zigzags around us. Bats have a very enhanced SOC because they effectively see with their ears when flying at high speed in the dark.

We don't have earlids to close; even when we are asleep, our hearing maintains its tireless vigil so that we don't miss our alarm call or a sound that might indicate danger to our unconscious body. All this goes on twenty-four hours a day, every day of our lives. Hearing never rests.

*

So far, we've looked at what happens when vibrations enter our ear, and how some sorts of vibration are processed by our brains. It's time to look at what all this means for your happiness, productivity and well-being.

2

HOW SOUND AFFECTS YOU

Most people don't know how to listen because the major part of their attention is taken up by thinking.

Eckhart Tolle

Take a moment to stop and listen to the sounds around you right now. The distant car alarm your brain had filtered out, the creak of a familiar floorboard, the wind in the branches of a tree, a distant neighbor laughing, the Doppler slide of an airplane as it passes overhead. So many different soundwaves, interrelating around you. Sound doesn't care at all that most of the time you are unaware of it: its effects still take place within you, often below the level of conscious thought.

Sound affects you constantly, in four powerful ways. Explaining this is the core of this book's purpose, because only when you understand these effects can you start to take responsibility for, and intentionally design, the sound you consume and the sound you create, dramatically improving your happiness, effectiveness and well-being—and that of the people around you. Sound is not an abstract, distant concept, or something to be suppressed; it's a potent force that shapes your enjoyment of, and outcomes in, life.

Sound affects humanity, and humanity affects sound. Simply by moving this powerful interrelationship to a conscious level—instead of ignoring it, as we have been for centuries—we can start to heal our individual and social relationships and respect the living things that share this planet with us. So, let's explore the four effects of sound.

1 Physiological

We have never outgrown our relationship with hearing as our first warning sense. Any sudden, unexpected sound will instantly create a fight/flight reaction in your body. Your heart rate will increase, and your body will release a cocktail of chemicals including epinephrine (adrenaline) and cortisol, which increase blood pressure and sugar and turn fatty acids into energy, preparing your muscles for immediate action. All this happens in milliseconds, while your cortex has only just started to assess the sound to decide if it is really a threat. Loud sudden sounds that occur with no visual warning—like an explosion, or even a fast train hurtling through a station—can actually kill as *in extremis* this fight/flight response can cause a fatal heart attack.

The typical resting human heart rate is around 60 to 80 beats per minute (bpm), but this can rise to over 200 bpm in states of extreme exertion, fear or arousal. It should be immediately obvious why music with a tempo below 80 bpm is typically perceived as calming, while 100 bpm is mildly stimulating and 150 bpm strongly exciting.

Entrainment is where two rhythmic processes interact with each other in such a way that they adjust toward one another and eventually "lock in." Later in the book we'll see how this impacts everything in the cosmos, but it also happens to you and me. The effect is often involuntary: for example, when up-tempo music causes us to walk faster in a mall or chew faster in a restaurant (both phenomena that have been repeatedly confirmed by research). If I transported you to a nightclub with 140 bpm dance music at high volume, your heart rate would immediately increase.

Sustained noise, even at the relatively moderate average level measured in classrooms in one German study, is bad for the health. It is now proven that prolonged exposure to this commonly encountered noise level increases a person's risk of heart attack and stroke, so teachers may well be shortening their lives by constantly working in noisy classrooms.

The auditory systems we encountered earlier are a miracle of biological precision engineering and they must last us a lifetime. Those bones of our ear weigh three thousandths of a gram and are constantly vibrating back and forth at frequencies up to many thousands of times a second. But it's those tiny hairs in our inner ears that are most vulnerable. Like our adult teeth, these are onetime gifts: once gone, they do not get replaced. The average adult has around fifteen thousand of them, a tiny number compared to the number of receptor cells involved in sight. If they get bent, they can't control their ion channels* properly. They may shut down, causing deafness, or they may fire continuously, causing tinnitus. If they get uprooted, there is no reforestation program. Excessively loud sounds will damage these delicate hairs, creating the cumulative degradation in our hearing known as noise-induced hearing loss (NIHL).

---

* The protein pathways expressed by any cell that allow charged ions to pass through the cell membrane.

Loss of high frequencies for a day or two after exposure to very loud sound (known as temporary threshold shift) is the first warning sign of hearing damage. If occasional, this is not anything to worry too much about, though that does depend on severity. In my teens, I saw Lynyrd Skynyrd play at London's legendary Hammersmith Odeon. The sound system was so enormous that I remember feeling sorry for the roadies, and their sound was appropriately huge. As I staggered out with the triple-soloing guitars screaming at the end of "Free Bird" still ringing in my ears, I realized I could not hear any sibilants (high-frequency consonants like s and t) at all when people spoke. This persisted all through the next day. A chunk of my hearing died that night. More was chipped away as I drummed at hundreds of gigs before I understood the dangers and bought myself some hearing protectors. As a result of the accumulated damage, I now can't hear certain higher frequencies, and have constant tinnitus with a tone ringing in both ears. Percussion is unsurprisingly a high-risk instrument, though research shows that brass and woodwind are equally dangerous—and, of course, with amplification any instrument, including the voice, can do serious damage.

With hearing, once the ice breaks, there is no going back. Studies suggest that almost half of all musicians suffer from noise-induced hearing loss—and not just those involved in heavy rock. A German study of three million people found that professional musicians were four times more likely to suffer from NIHL than non-musicians, while analysis from the Chicago Symphony Orchestra found that even violinists were damaging the hearing in their own left ears.

Of course, there are many other jobs that put hearing at risk. Construction, mining and any industry using large machinery are all legally obliged to take great care and offer workers hearing protection, but many employers, and workers, ignore the advice. A 2005 paper estimated that occupational NIHL accounted for around a sixth of all global hearing loss.

Even if we don't expose our ears to excessive noise, age and illness take their toll on hearing. There is simple wear and tear: as we get older, the hairs become more brittle and less effective, even if they are undamaged by outside agencies, and the result is known as age-related hearing loss, or presbycusis. Also, diseases such as meningitis and rubella, or any inflammation of the middle ear, can produce severe hearing loss. In addition to permanent damage, common ailments like colds, wax build-up or glue ear temporarily affect millions of people's hearing every day, especially children. It's often forgotten in designing schoolroom acoustics that, on any given day, a sizable section of any group of children will probably be struggling to hear well.

Around a quarter of the population of the world's industrialized countries have a hearing problem, mainly simply due to age. By the age of fifty, around twenty percent of people have some form of hearing impairment; by seventy, the proportion is more than half. The World Health Organization (WHO) estimates that over 1.5 billion people in the world currently live with hearing loss, and predicts that this will rise to 2.5 billion by 2050. That will be one in four people on the planet.

In even this huge number, I'm not sure the WHO is yet factoring in what I think is the biggest hearing-loss time bomb: headphones. A recent global study estimated that over a billion young people are heading for NIHL because of listening to music on headphones at excessive volume—something the study revealed that one in four young people were doing. Any time you can hear the music leaking out of someone's headphones, they are harming their hearing. Doing this for hours a day is a recipe for serious hearing damage that will not be evident for years, in most cases, so there are no immediate symptoms to warn the headphone abusers. We could be dealing with a largely deaf generation in twenty years' time.

Age tends to erode not only hearing, but also listening. The degradation of the cocktail party effect is a well-known example: I certainly find it harder to extract signal from noise in a crowd than I once did, and that's only partly to do with hearing loss. Perhaps this is in line with declining cognitive ability in general, as capacities like focus, memory and learning tend to wane with the years. Becoming more conscious of listening by using the exercises in this book is a wonderful way to combat such decline because it exercises those mental muscles continuously, as well as often revealing delights (or dangers) as one savors the sounds around.

Severe hearing damage has a massive effect on quality of life: removing the ability to enjoy music or birdsong is a profound loss, and, of course, there are safety implications too because hearing is our primary warning sense. But, most of all, this matters because losing the top end of the frequency spectrum, which is the most common form of hearing loss, affects speech intelligibility. In nontonal languages, we understand speech largely through consonants, as can easily be appreciated by considering the following, where all the consonants have been removed: *e ou e eai i a ae ea eeaio i e ea.*

Now try again with all the vowels removed: *w cld b dlng wth lrgly df gnrtn n twnty yrs.*

Higher-frequency consonants include s, t, f, p, k, z, th, sh, zh, h, v and d. When you shout, you increase the volume of vowels far more than consonants, which is why it's harder to understand someone who is shouting from a distance. People with damaged hearing typically lose higher frequencies so they experience exactly that problem all the time, which can occur as a humiliating handicap that many people try to disguise, instead of seeking help. Untreated hearing loss often leads to damaged relationships, which can turn into isolation, depression and even suicide.

Along with the quality of life, relationship and mental health issues, hearing loss has a huge societal cost in loss of productivity and education: in most occupations, hearing well is a necessary capability, and it is equally vital in most learning institutions. The WHO's *World Report on Hearing* estimates the cost of unaddressed hearing loss at $980 billion a year, and the return on investment in ear and hearing care, especially in low- and middle-income countries, to be $16 for every dollar spent.

In many cases, there are now effective remedies. Hearing aids have come a long way in the last couple of decades. I've worked with some of the leaders in the field, and this is one area where technology is unreservedly beneficial: modern devices are discreet, powerful and very smart, with apps allowing them to be configured for different contexts and individual needs. The quality of the sound is far better now than it was in the twentieth century, even allowing the enjoyment of music. For those who seek help, it is now available, albeit at a cost in both money and pride. Sadly, this combination of obstacles results in less than one in five of those who would benefit from a hearing aid using one, according to the WHO.

In the future, there may be cures using stem cells to regrow damaged hair cells or auditory neurons. There are well-funded companies working on this at the time of writing, such as Rinri Therapeutics in the UK and Lineage Cell Therapeutics in the U.S. This is a prize worth pursuing: as Rinri's CEO Simon Chandler says, "It is well recognized that hearing loss significantly impacts quality of life, affecting individuals through to economies. A biological solution that can restore hearing to a significant level is potentially transformative to patients." However, achieving this is a long-term project: hearing loss is complex, and so is the task of replacing failed cells in the inner ear. I doubt we'll see any magic bullets for hearing loss in my lifetime.

Some people suffer from the opposite of hearing loss. Hyperacusis is a condition where everyday sounds can occur as unbearably loud, sometimes painful. A similar but distinct condition is misophonia, where the sufferer feels intense anger, panic or upset on hearing trigger sounds, which are often mundane human-generated noises such as eating or breathing. Hyperacusis can occur after an injury or Lyme disease, while misophonia typically appears in teenage years, seems to be neurological and, while usually auditory, can also be triggered by visual stimuli: I once had a colleague who was deeply disturbed by the sight of buttons! For a fascinating exploration of a range of even more exotic sound- (mainly music-) related conditions, I recommend the entrancing book *Musicophilia* by the late neurologist and psychiatrist Oliver Sacks.

WHAT YOU CAN DO

Avoid unnecessary sudden noises. Old-fashioned alarm clocks wake people with an instant fight/flight response, which is far from the best way to start your day. Replace an alarm bell with birdsong, ideally with a gradual fade-in peaking at your waking time.

At the other end of the day, if you have trouble getting to sleep, listen to gentle surf sounds, which have similar rhythms to the breathing of a sleeping human and are generally associated with peace, relaxation and beauty.

Get a basic noise meter on your phone or other device—some already have this built in. Use it to alert you when you are in excessively noisy spaces and, unless that's why you're there (a gig or a nightclub, for instance), move yourself to the quietest location or even leave if you can. If it's a restaurant or bar, tell them why you're leaving! Most people don't, so the noise continues with the managers oblivious to the discomfort it causes. If it's your place of work, raise the issue and seek solutions like installing acoustic treatments or removing noise sources (more on these later).

Be careful with headphones. As a rule of thumb, if you can't hear someone speaking to you in a loud voice from around a meter away, you probably have the volume too loud for extended listening. Get the best headphones you can afford because cheap headphones, like cheap loudspeakers, constantly tempt you to push the volume up in order to make them deliver acceptable frequency response, particularly bass. Most of all, teach your children about the danger of hearing damage and responsible listening habits.

2 Psychological

Sound affects not just our bodies but also our emotions, moods and states. Music is the most obvious example: it can energize and excite, soothe and relax, or generate almost any human emotion you can think of. But music is not the only sound that affects our feelings. Birdsong is a wondrous sound that's been on Earth far longer than human beings have, and one that we've learned through the millennia to associate with two useful states. First, birdsong is nature's alarm clock, so we associate it with being awake, and it helps to stimulate alertness. Second, when the birds are happily singing things are normally safe, so birdsong makes people feel secure: by contrast, there are many reports of birds falling silent before tsunami waves or earthquakes strike or when a predator appears, so a sudden silence can be discomfiting.

Birdsong can be an important element of biophilia, a movement in interior design that connects us with nature in spaces which may be closed off, grimly artificial or overly austere. In my work with organizational sound, I have deployed birdsong in shops, airports and even service station toilets, with excellent results for these reasons; it's also a useful sound for workspaces, where a mind-alert/body-relaxed state is ideal.

Association is very important in determining both the impact of all sound on human beings—the effects of any given sound can be very personal, depending on one's accumulated life experiences—and the resulting memories and connections. In his biography, the footballer Wayne Rooney reveals that he can't sleep without the sound of a vacuum cleaner, presumably after his mother used sounds like that to help him sleep when he was a baby—or maybe she just did the cleaning after making him comfortable. These kinds of associations show how a simple sound can instantly transport us decades back into a past experience, and sometimes move us deeply as a result.

WHAT YOU CAN DO

Take some time to think about the sounds that affect your feelings and the associations behind them, then write them down in two lists: positive and negative. Find ways to surround yourself with the positive ones at appropriate times, just as sports stars will listen to particular music to help them achieve peak performance. Equally, plan to reduce the instances of the negative sounds.

3 Cognitive

If you've ever tried to write, calculate or think rigorously in a noisy environment, particularly one where people are speaking around you, you know how distracting sound can be. You have no earlids, and evolution has programmed you to pay attention to human voices and to decode language. A substantial body of research now reveals how damaging open-plan offices are for solo working, for exactly these reasons. The papers vary in quantifying the effect; the most startling is a 1998 paper by psychologists Simon Banbury and Dianne Berry which found that the sound of a noisy office reduced people's productivity in manipulating words or numbers by a staggering two-thirds.

Research into the irrelevant sound effect shows the top three features that make any sound more distracting: variability, density and association. You can habituate to a constant noise like a humming fridge or hissing air conditioner (though it still takes some mental work and will be tiring over long periods), but a door that bangs shut every few minutes is a different matter. Dense sound (that's sound with a large number of events or changes in a short time) contains a lot of information, and is much more attention-grabbing than simpler, sparser sound. And if a sound is associated with the need for action, it's hard to ignore it. We are genetically programmed to care for babies, so the sound of a baby crying is disturbing, especially if it's not your baby and you can't do anything about it, as many a traveler on a long-distance red-eye flight will testify; the same applies to alarms or ringing phones.

Unwanted conversation is the ultimate distraction: it's variable, dense and we have a primal inner imperative telling us that it must be listened to. I was lucky enough to see Mihály Csíkszentmihályi give his excellent 2004 TED talk on his concept of flow state. In that talk, he said: "Our nervous system is incapable of processing more than about 110 bits of information per second. And in order to hear me and understand what I'm saying, you need to process about 60 bits per second." This is why you can't understand two people speaking at once, while someone speaking when you're trying to think is taking up much of the bandwidth you need in order to listen to the inner voice you use when you process words or numbers, hence the dramatic drop in efficiency.

WHAT YOU CAN DO

If you have to work in a noisy open-plan office, I suggest you take an MBA—not a business degree, but a little mnemonic that may help.

M stands for move. Many modern offices are now creating quiet working spaces, and some have adopted a strategy known as activity-based working (ABW) where they offer a variety of workspaces and encourage people to move around and choose the space that best supports the work they want to do.

B stands for block. If you can't move, or if there are no suitable spaces available, you can block the distracting sounds by using headphones. But please be careful about what you play: for most people, music is just as distracting as unwanted conversation, albeit much more enjoyable. Unless you have really established that the music you choose helps you to work more effectively, you might do better with a less distracting sound, such as birdsong or other gentle nature-based sounds. I suggest you test various sounds to see which suits you best, as we're all different: I'm sure there's someone in the world who can think at their best to the accompaniment of deafening death metal music, but that's definitely not me. Some years ago, I worked with acoustics specialist Ecophon to create an app called Study, which is still available at no charge in the usual places. It plays forty-five minutes of sound that's specifically designed to enhance cognition, then stops to remind you to take a rest. These days, if you search on YouTube you can find hundreds of alternatives, so try some, see what works best for you, and then have your working sound and your headphones always at the ready to block out distractions.

A stands for accept. If you can't move and you don't have the means to block, raging at the distractions is very tempting—but this in itself will become potentially the biggest block to your productivity. The best strategy is to accept the noise and tell yourself that you'll just do your best in the circumstances, without the grumbling.

4 Behavioral

Sound can create social behavior. Groups of human beings exhibit entrainment and sync in behaviors such as clapping at the end of a performance, where people tend to fall into unison, or walking, as famously reported when the London Millennium Footbridge opened in June 2000 to link St. Paul's Cathedral with the Tate Modern gallery across the Thames. As soon as people started to walk across it, the bridge started swaying alarmingly from side to side, to the point where people were feeling sick and hanging onto the sides in panic. The bridge had to be closed immediately while engineers worked out what was happening. They found that the bridge had a "lateral frequency mode"—a tendency to vibrate horizontally at around one cycle per second, which happens to be about half the typical frequency of human footfall. What the designers had not modeled was an active feedback loop between the bridge and the people crossing it. When the bridge moved, everyone on it reacted by taking a sideways step at the same time, pushing the bridge in the opposite direction and amplifying the lateral movement. The Millennium Footbridge eventually reopened after two years of remedial work.

Sound powerfully affects how you and I behave on our own, too. In a famous 1982 paper, professor of marketing Ronald Milliman found that playing slow-paced music in a supermarket caused people to walk more slowly, so they stayed in the store seventeen percent longer than when fast-paced music was played. The startling result of this was that sales rose by thirty-eight percent, presumably as slower-moving people noticed more things on the shelves that attracted them.

Many other aspects of sound can affect what we do and how long we stay in a place. At the most basic level, we will move away from unpleasant sound if we can—something that's widely ignored by retailers who often blast customers with inappropriate music through poor quality sound systems. The type of music played can be significant, as outlined in many research papers: Yalch and Spangenberg in 2000 found that familiar Top Forty pop music led to shorter shopping times, while Areni and Kim in 1993 found that in a wine shop classical music caused people to choose more expensive bottles, probably because we associate classical music with quality. Guéguen, Le Guellec and Jacob in 2004 found that loud music in a bar caused people to buy more drinks, so the demise of the quiet pub or bar sadly seems justified commercially, if not aesthetically.

To me, the most startling study on sound and behavior was carried out by Adrian North and his colleagues in 1999. They alternated French and German music in a supermarket and tracked the sales of French and German wine, which were both displayed in identical fashion. On French music days, French wine sold five times more than German, which may not be surprising as it does sell more globally. But—and this is a huge but—on German music days, German wine outsold French by two to one. This massive shift in people's purchasing behavior was completely unconscious: when asked, most people had not even noticed the music.

If that's how powerfully sound can affect what you do and the choices you make even when you don't pay attention to it, it makes sense to start listening consciously so that you can reclaim dominion over your actions instead of unconsciously responding to auditory cues. Meanwhile, businesses stand to lose or gain so much from the quality and appropriateness of their sound that it's baffling to me that so many of them give sound such scant thought, if any at all. Designing soundscapes is just as important as designing how things look, but in marketing and retail, as in most things, the eyes have it, and the ears are ignored.

WHAT YOU CAN DO

Develop the habit of attending to the sounds around you, like a musician listening to the rest of the orchestra. Now that you know the main behavioral drivers—tempo, liking, volume and association—you can start to become conscious of the effects of each soundscape on your behavior, up to and including how you spend your money.

For organizations: carefully consider the soundscapes you are imposing on staff, customers and people in general in spaces you control, from offices and factories to shops, showrooms and other public buildings such as malls or places of worship. Think about all four effects of sound and ask yourself: is this the ideal soundscape for the benefit of the people in this space and for the outcomes we are trying to achieve?

Crossmodal effects

There's one more complexity to consider. Definitions vary, but most academics agree that we have something between five and ten senses, and it is now established that they constantly affect one another in what are known as crossmodal effects.

A leading specialist in this field is Charles Spence, Professor of Experimental Psychology and head of Oxford University's Crossmodal

Research Laboratory, who has written several fascinating books on the subject. I spoke to Charles at his house in the hills above Bogotá, Colombia, and he explained that the eyes take priority: "You can fool wine experts into smelling red wine aromas simply by coloring a white wine red, because we're all visually dominant." Another key process that determines our reality is "superadditivity"—where the senses are all given congruent signals and an experience has a much greater impact as a result. It's opposite, which I'm pleased to say the scientists do not call "super-subtractivity," when the senses are given conflicting signals, results in a suppression effect.

The famous McGurk illusion [1] illustrates the power of cross-modal effects: a video of a face saying "ga ga" has a soundtrack of the same person saying "ba ba." Look at the face and you will almost certainly hear "da da," as the visual information clashes with the aural: close your eyes, and "ba ba" can be clearly heard.

Clearly, sight overrides sound here—but Charles Spence has discovered that sound can equally affect the other senses in mysterious ways. For example, he told me: "Airplane noise seems to selectively suppress our ability to taste sugar and salt in airline food, but actually enhances our ability to detect and perceive umami, the mysterious fifth taste, perhaps explaining why so many airplane passengers drink Bloody Marys or tomato juice on the airplane when they never do so on the ground."

Brand and product designers are starting to spend heavily on seeking superadditivity. Car manufacturers have been doing it the longest, carefully padding hollow metal doors so that they make a satisfying, luxurious "clunk" when they close—far from their natural sound, and a sort of consensual myth that buyers and manufacturers have co-created. I asked Charles Spence what kind of products his lab has worked on, and he replied, "Everything: bags of crisps, bottles, cans, corks, stoppers, any kind of closure, any kind of device that makes a sound—and the scope to engineer it to better match the desired, expected or hoped for brand attributes." That new car smell? Completely artificial, carefully designed and sprayed in before sale.

Notwithstanding such learned associations, Spence believes there are primal underlying correspondences. For example, the bouba/kiki effect (where bouba is associated with a soft, curvy shape and kiki with a sharp, angular one) is universal, regardless of language. Spence's research has found equally universal associations of sweet taste with roundness, light and high pitch, while bitter taste is associated with dark, low pitch and angularity. There are, of course, many other correspondences that arise from local culture, such as the very different associations with colors: in the West, white generally means purity and peace and is used in weddings, whereas in the East it can mean death and sadness and is used in funerals; similarly, red means danger in the West, but luck and prosperity in the East.

I asked Charles for his tips on using crossmodal effects for a healthier, happier life. His tip was to bring into our homes and workplaces the sights, smells and sounds of nature: "Trying to think more carefully about the way we evolved might put us in a better position to design the multi-sensory environments where we live, where we work, where we play, where we sleep, so that they more optimally stimulate our senses in a congruent manner, and in the manner that we evolved for—and which will then hopefully have the beneficial outcomes that so many of us are in desperate need of these days."

We have already briefly encountered the recent trend to push back against sterile spaces and reconnect with the natural world known as biophilia, and it's something I strongly believe in. In 2022, I was thrilled to be involved in the launch of a company called Moodsonic that creates beautiful nature-based soundscapes to improve both productivity and well-being in workspaces, each one generated live by computers so they don't repeat themselves, just like natural sound. I hope that in future we will all be able to reforge our links with nature in all our senses, everywhere we go, because we can already see all too well the devastating outcomes that arise when we shut ourselves off and start to see the planet as just a resource for more.

Designing with your ears

There are ways you can modify the sound in your space to make you happier, healthier and more productive. If possible, your first step should be to improve the acoustics, which are all too often ignored by architects. To counteract unwanted reflection and transmission, we can use the ABC of acoustics: absorb, block, cover.

The surfaces so popular in modern interior design—glass, metal, stone, plasterboard; wood too—bounce back most of the sound that hits them, making a room noisier and more stressful to be in, as well as compromising communication, because unwanted reflections often mask or confuse the sound of someone speaking. Sometimes this can make a room deeply unpleasant because of standing waves.

Diffusors can help to reduce the nastier aspects of reflection: they are essentially just uneven surfaces that effectively smash the soundwaves and spread the sound evenly around, which is usually much more pleasant for anyone in the space. The trick is to avoid rooms with hard, featureless parallel walls—just a bookcase or some plants can help enormously.

Soft surfaces like carpets, curtains, drapes, plants and soft furnishings all soak up sound, making spaces quieter. Until recently such absorbents were often drab and utilitarian—square white ceiling tiles in a grid, or bulky industrial-looking panels on walls. From my work over the years with Armstrong Ceilings in the USA, I have discovered to my great delight that we now have a new wave of treatments that can both absorb and block sound—and look superb at the same time. No longer is there a trade-off between pleasant sound and good visual design: today's wall panels can carry any graphic, color, picture or artwork and still soak up sound; suspended baffles and clouds can be any shape and color; even beautiful wood and metal ceilings or walls can do a great job if you perforate them and put absorbing material behind them; and there are effective acoustic ceilings that look like standard plasterboard—and we now even have excellent acoustic plaster as well.

Sometimes you will need to add the acoustician's C: covering unwanted sound with some other sound. This is known as masking. It's usually deployed when unwanted speech is distracting people or compromising privacy in an open-plan space. Masking will not work to hide high levels of unpleasant noise, but it can blur conversation, making it harder to understand people a few meters away while you can still focus on the person next to you; it can also hide unpleasant background noise at low level, such as traffic noise outside the room.

Air-conditioning equipment can create some masking with its hums and hisses, or a dedicated masking system can be installed. Old-fashioned ones use a sound that's not much more pleasant than air-conditioning—filtered mechanical noise that sounds like a constant low-pitched hiss. Because you will habituate to any constant sound, you will cease to hear such a sound after a while: your brain will say, "I know what that's doing so I'm not listening to it." However, this is still mental work, as you will remember if you've ever been in an office at 6 p.m. when the air-conditioning shuts off and suddenly everyone's shoulders relax. Masking noise with more noise may increase productivity, but it also increases stress.

Another option is music. At a moderate level, it can work well for masking—but only if people are not trying to concentrate on solo working, if everyone in the space agrees to its use, and if the music is not overly attention-demanding. There is a key proviso here: music on top of unpleasant noise is simply more noise, just as perfume on an unpleasant smell simply makes a worse smell. This is a scenario you may have encountered in coffee bars or restaurants, where bad acoustics, noisy furnishings, banging baristas and shouted conversations compete with semi-intelligible music that's making nobody happy.

By now, you will be unsurprised to learn that research suggests that the best options for masking are the natural sounds that so many people miss now that most of us live in cities and spend ninety-five percent of our lives indoors. Wind, water and birdsong are proven to be good for us, reducing stress and creating positive emotions. They can also mask conversation—especially water, because it occupies the same frequencies as the sibilant consonants (mainly s and t) that are so key to comprehending speech. Try talking to someone next to a waterfall and you'll experience how effective water is at blurring language. This is exactly why I was excited to help with the birth of Moodsonic, which is now used by major organizations worldwide to create health-giving, productive sound in offices and hospitals. I'm sure others will follow its lead, and, in years to come, many indoor spaces will feature the pleasant and nourishing sounds of the natural world to go along with other forms of biophilia for the eyes and the nose.

Sound that you enjoy can cause the release of "happy hormones" like endorphins, dopamine and oxytocin, as well as the immunity-boosting antibody secretory Immunoglobulin A. Pleasing, calming sound has been shown to reduce blood pressure and stress (for example, before a medical procedure) and even reduce pain. Explore what's pleasing for you and make a point of surrounding yourself with it whenever possible.

WHAT YOU CAN DO

How can you tell if the sound in your home or workplace is working against you? There are two simple measurements that you can use to make sure any space is fit for purpose.

First, noise level. A good intuitive test is to ask yourself: how loudly would I need to speak to be well heard in this space? If the answer is shouting, you're into noise levels that are potentially dangerous to health in the long run. There may be what's called a noise floor in the room—for instance, where there's electromechanical noise from air-conditioning, heating or computer fans. A high noise floor will automatically make a room louder when people enter and start talking over the top of the existing sound. If you can, try to minimize the noise floor, possibly by turning down or moving equipment.

Second, reverberation time (RT). Technically, this is the time it takes for a sudden sound to die away, usually by 60 dB. You can get a good idea by simply clapping your hands and counting the seconds until you can't hear anything. In a cathedral, that might be anything up to ten seconds, which is great for the plainsong but not so useful for a rock gig or a meeting. For an open-plan office, anything around one second is reasonable in my experience. For a meeting room or classroom, RT should be well under a second, and ideally less than half that.

These are very crude measures, but they're a great start. Once you've improved the acoustics, think about the best sound you can inject into the space, with nature sounds at the top of the list. Experiment until you get a result that works for you or your team.

Sonic healing

Another, much more ancient connection between sound and our bodies concerns the idea that sound can heal illness or enhance health. Like the medicinal use of plants, sound-healing dates back thousands of years—and, in similar fashion, it has been elbowed out of the mainstream by being labeled as witchcraft by religious bigots or dismissed as a crank notion by conservative rationalists. But just as we derive many modern drugs from the active chemicals in the plants that ancient healers knew so well, so modern medicine is now finding that sound can powerfully help to fight illness.

The practice of using sound to heal almost certainly goes back as far as humanity, in the form of chanting or other vocal interventions. Instrumental music has been used in therapy for at least four thousand years: in 2000 BCE, the cuneiform writings of Assyrians depict the use of music to deflect the path of evil spirits. There are records of music being used intentionally for health in ancient Egypt and, of course, in Greece, where Pythagoras taught that harmony is health, and that music is a core element of science. Aristotle and Plato continued the association between music and well-being, though they both focused on the emotional, political and spiritual benefits rather than physical ones.

There is now a mass of research on the therapeutic use of music, and the American Music Therapy Association is a good place to access much of it. The evidence shows that music can help with many conditions, including dementia, autism, addiction, post-traumatic stress disorder and stress and pain management. Many surgeons, for their own purposes, play music while operating, and still more offer patients the option to listen to their own selections before, during and after the procedure; the evidence shows that this can reduce anxiety and even the amount of anesthetic required.

There is far less research on the efficacy of sung mantras, chant or tonal sound-healing using singing bowls, tuning forks, bells or gongs. On average, a human body is sixty percent water, and, as we have discovered, water is an excellent medium for sound. You are in effect a chord, with every part of you vibrating, from subatomic strings to atoms to molecules to cells and organs, so it doesn't seem too far-fetched to suggest that one definition of good health is harmony in that chord, or that dissonant elements might be entrained back into harmony by external sounds. I have certainly felt the emotional and spiritual benefits of these methods, and their long history would indicate that many other humans have too, but whether external sound can foster deep healing and cure serious conditions is as yet unproven and very under-researched. I admit to a strong suspicion that, as with herbal remedies, science will, in time, rediscover that there is no smoke without fire, and that the ancients knew what they were doing.

There is one other powerful therapeutic use of sound: as we've already found, playing music is definitely good for you. Learning an instrument can improve reaction times that slow down naturally with age, help with integrating sensory inputs, and stimulate the brain. If you have not yet made any music in your life, it is never too late to start.

There is one qualifier: we are talking here about casual, amateur playing. The life of a professional musician is not always a healthy one, and even amateur bands and orchestras can be loud enough to damage their members' hearing.

WHAT YOU CAN DO

Sound can have a profound effect on your well-being, for good or ill. I suggest you spend a few minutes to ask yourself: what sounds do I surround myself with at home, at work, or in my leisure or hobbies, and how could I change them to enhance my health? The answers may surprise you.

The sound of music

Intention is very important for all sound. Almost all music is made to be listened to and is both information-dense and fast-changing, especially if it includes vocals—the rare exceptions being new age, spa or ambient music, which are designed to be sparse and slow-changing with the specific intention of being relaxing, or perhaps supporting working.

Dense, changeable music, whether folk, pop, rock, jazz, urban, dance or classical, demands a lot of attention, and, for most people, that demand will reduce their productivity. If your children insist that they can do their homework better with music on, there may be some truth in it: they may become more willing to do the work, which might be a step forward; also, they may keep at it for longer as they will have a more enjoyable experience, so it may be that the task will get accomplished more often. But they will almost certainly not get more done per minute, because the music will take up a chunk of their cognitive processing.

This generalization needs to be tempered by recognizing that everyone is different. Not all tasks are the same: obviously, boring, repetitive work can be lightened by very attention-grabbing music that takes you away from the task and its tedium. Nobody has the right to tell everyone else what will work best for them in sound.

WHAT YOU CAN DO

To enhance your productivity with sound, the best approach is to try different sounds and see which ones help you get the most done, whether that's in terms of motivation, clarity of thought or staying power. As I've already suggested, please do try nature sounds, because birdsong and gentle water sounds (bubbling streams, not surf, which is soporific) are good at masking conversation and yet very easy to ignore because they have formed the backdrop to human activity for countless millennia.

Now that you know how sound can affect your happiness, effectiveness, and well-being, you might like to explore a little more deeply what sound really is. That is the subject of our next chapter.

3

SOUND UNDERSTANDING

I like to listen. I have learned a great deal from listening carefully. Most people never listen.

Ernest Hemingway

"Just start singing," said David. "Don't worry about whether it's a note, or where it's going or what it sounds like. No plan. Just sing!"

I was standing with a group of twelve people in the remains of the twelfth-century Højerup church at Stevns Klint on the eastern coast of Denmark. (It was the remains of a church because a large chunk of it had slid down the chalk cliffs into the sea below several years before, hence the lovely space we were in was deconsecrated.) As a drummer, I was never a jazz player, and I always liked to understand the structures of songs in the rock bands I played with—to know each story's ending in advance—so it was a little intimidating for me as we stood in a circle eyeing each other. We were under the expert guidance of David Hykes, the American overtone singer, composer and author, who studied under Gurdjieff, Sheila Dhar and the Dalai Lama. I had come across the extraordinary 1983 album by David's Harmonic Choir group, *Hearing Solar Winds*, which was unlike anything I'd ever heard and haunted me for weeks. When I discovered that David gave workshops to teach the exquisite harmonic chant techniques he had developed from the Tuvan and Tibetan traditions he'd studied extensively first-hand, I jumped at the chance. Over the next few days, we learned to focus overtones by changing the shape of our mouths as we sang, and I started to hear the harmonics in every sound, rather like seeing the rainbow spectrum of colors in every light. It was as if I had switched on a sensitivity that had lain dormant all my life. When I started my rental car, instead of a dull rumble I heard the full glory of all the harmonics in the engine sound.

But what was happening?

*

We've established that sound is audible vibration, and described what happens once the vibrations reach our ear. But what is vibration and how can we understand and describe it?

My old childhood favorite the *Encyclopaedia Britannica* defines vibration as "the periodic back-and-forth motion of the particles of an elastic body or medium, commonly resulting when almost any physical system is displaced from its equilibrium condition and allowed to respond to the forces that tend to restore equilibrium." If regular or repetitive, like a pendulum, we call this movement about a central point oscillation; vibration is a wider term that includes irregular back-and-forth motion of an object or system—for example, an earthquake.

Everything vibrates, from the tiny to the vast: inside the cells in your body right now, molecules are vibrating nanometers at a time over millionths of a second—while in 2006, a team from the University of California, Berkeley, led by professor of astronomy Leo Blitz, discovered that two dwarf galaxies are causing the entire gas disc of the Milky Way to vibrate in three distinct notes. As the Magellanic Clouds force their way through the dark matter halo that extends far beyond the stars in our galaxy, they are exciting a resonance at its center which in turn causes an oscillation in the gas disc embedded in the halo. "It looks like the Milky Way is flapping in the breeze," commented Blitz. The team was able to isolate three different types of vibration about sixty-four octaves below middle C. That's an extremely low frequency of one complete oscillation every 2.2 billion years.

In between these extremes is a huge spectrum of speeds of oscillation. When an oscillation propagates from its source, typically through a medium like air or water, we call that a wave. We should note that not all types of wave need a medium to propagate. Electromagnetic waves such as radio, microwaves, infrared, visible light, ultraviolet, x-rays and gamma rays move through something known as the electromagnetic field (EMF) and can travel through a vacuum.

But sound is what is known as a mechanical wave, and it does need a medium to move through. That can be a solid, liquid or gas. Most of the sound we experience is propagated through air, but it's easier to imagine soundwaves as the ripples that result when a stone is dropped in a pool of water. When our tree in the forest falls and releases some of its kinetic energy as a vibration, it bumps the adjacent air molecules, which in turn bump into their neighboring molecules, spreading outwards in a circular pattern just like the ripples in that pond—though as we live in a three-dimensional world, the soundwaves are spherical, not just ripples on a horizontal surface.

The speed of sound

The speed of soundwaves is not a constant: it depends on the medium through which the sound is traveling. The stiffer the material or the lower its density, the faster sound travels. In air, at room temperature, soundwaves travel at about 340 meters per second (mps), which is just over 1,200 kilometers per hour (kph). At this speed, known as Mach 1, sound covers a kilometer in almost exactly three seconds, which is why it's easy to calculate the distance of a lightning strike from you in kilometers: just count the seconds to the thunder and then divide by three.

Many people are surprised to learn that water is a far better medium for sound than air. This is because it's far less compressible (stiffer), so sound travels over four times faster in water than in air, at around 1,500 mps (5,400 kph). Metals are stiffer still, so they transmit sound even faster. In iron, sound travels at over 5,000 mps (18,000 kph), and in an exceptionally stiff material such as diamond, sound can reach 12,000 mps (just over 43,000 kph), about thirty-five times its speed in air.

As the noise of our falling tree propagates, air molecules are bumped out of their position before coming back to where they were, creating a series of waves. We can describe these waves using three basic concepts, all of which impact how we perceive a sound and help us to describe that experience.

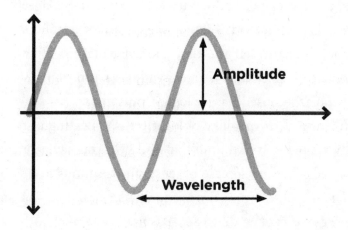

Amplitude

Our first basic concept is amplitude—the maximum disturbance of a wave from its undisturbed position or equilibrium. We experience this as the loudness of a sound: more energy moving molecules further from their equilibrium is perceived as a louder noise. A wave carrying greater energy vibrates our eardrum more and creates larger oscillations when it reaches our cochlea. This is why a leaf is quieter than a plate when it hits the ground. Like those ripples in the pond, soundwaves dissipate with distance, which is why noises further away seem quieter.

Sound pressure is measured in decibels. Normally we measure pressure in pascals, which are defined as one newton per square meter. But we rarely use pascals to describe sound pressure level (SPL) because human beings have a huge range of perception: the quietest sound a person could hear is defined as 20 micropascals (20 µPa), while a rocket launch (which would instantly burst your eardrums should you be unwise enough to be close by) is around 20 billion µPa.

A scale with a range of a billion to one is not very manageable, so we use decibels (dB) instead. These are named after telephone inventor Alexander Graham Bell and they condense that range into a much more practical spectrum of zero to 180—and in doing so confuse many people, because they are logarithmic.

If, like me, your memory of logarithms is fleeting at best, there is a very simple rule of thumb that will help you navigate your way around decibels. An increase of ten decibels sounds to human ears like a *doubling* of volume; a drop of 10 dB like a halving. Thus, moving from a noise level of 60 dB to 80 dB is not just an increase of a third: it's four times as loud. This is what most people don't understand: they confuse the logarithmic scale with a linear one, which may be one reason why many places are so dangerously noisy.

Below is a table of decibel levels for some relatable sounds to give you some reference points. We'll be using decibels often from here on, especially when discussing noise. This table combines data from several sources, including a much-quoted 2004 table created by Californian physicist William Hamby [2]. In our table, decibel levels are A-weighted and normalized to a distance of one meter from the source, unless otherwise stated.*

If tables are not your thing, some useful benchmarks to remember are a quiet bedroom at 30 dB, normal conversation at 50 to 60 dB, noisy traffic at 80 dB, and pain with permanent hearing damage at 120 dB. Strictly speaking, 194 dB is the loudest possible sound in

---

* A-weighting allows for the different sensitivity of the human ear at various frequencies, and distance must be standardized because sound pressure level falls as distance increases, in accordance with the inverse square law, so without a distance attached a decibel reading is virtually meaningless.

air. Anything beyond that is really a shockwave, where the pressure pushes air in front of it rather than compressing and rarefying the air molecules in waves. Shockwaves move much faster than the speed of sound, and degrade quickly over distance, fading to become sound as they do so. However, it's interesting to carry our scale up in order to get an idea of where the loudest events are placed. And, always remember: something 10 dB higher has ten times the sound energy and sounds twice as loud; 20 dB higher means one hundred times the energy, sounding four times as loud.

dB	Description
-21	Microsoft anechoic chamber in Redmond, Washington State, the world's quietest place
0	Threshold of hearing: a mosquito at three meters. Pressure variations are less than one billionth of atmospheric pressure; air vibrations around a tenth of an atomic diameter
10	Normal breathing
20	Empty concert hall/rustling leaves
30	Quiet bedroom at night/soft whisper
40	Library
50	(Ambient) Office, moderate noise level/moderate rainfall
50–60	Normal conversation
65	(Ambient) Level at which long-term frequent exposure increases risk of heart disease
70	Car interior/vacuum cleaner at three meters/busy restaurant
80	Heavy traffic at ten meters
85	Beginning of hearing damage: earplugs should be worn for long exposure and are specified by law in many countries
90	Heavy truck at ten meters/lawnmower or motorcycle
100	Typical car or house stereo at maximum volume/loud shout/jackhammer
110	Nightclub on the dancefloor/loud thunder heard at ground level/chainsaw
115	Permanent hearing damage within a few minutes
120	Threshold of pain/front row at a typical rock concert/jet taking off at 50 meters

130	Jackhammer or machine gun at ten meters
140	Gunshot at one meter—instant damage to hearing
145	Human nose itches/vision blurs
147	Formula One racing car
150	1976 rock concert by The Who, normalized to one meter. (Manowar and Kiss concerts may have been even louder.)
152	Painful vibration felt in all joints; it is almost impossible to swallow
160	Drag racing cars: 5,000 to 7,000 horsepower, liquid nitromethane fuel, earth-shaking at fifteen meters/humans find it hard to see and breathe
163	Whalesong
150	Fireworks at professional shows
165	Passenger jet taking off
170	World's loudest car stereos, up to eighty speakers, thirty-two car batteries, 100 kW
190	Eardrums rupture fifty percent of the time
198	Death from sound (shock) wave alone
204	Saturn V rocket taking off (NASA estimate)
212	Sonic boom from jet
215	Battleship firing nine 40 cm guns
229	Seafloor volcanic eruption
235	Earthquake at 5.0 on Richter scale
240	Tornado with 300 mph wind
278	Bikini Atoll 1954 nuclear test—15 megatons
282	Russian Tsar Bomba test: largest hydrogen bomb ever detonated (50 megatons). Shockwaves circled the Earth three times
286	Mount St. Helens volcano eruption blew down trees twenty-five kilometers away and blew out some windows in Seattle, 320 kilometers away
296	Earthquake 8.6 on Richter scale—ground moved up and down thirteen feet
302	Tunguska meteor, Siberia, knocked people over twenty kilometers away
310	Krakatoa volcano eruption, 1883, heard 4,700 kilometers away; this sound covered ten percent of the world's surface

Rather confusingly, the decibel scale is different for underwater sound. There are two reasons for this. First, while zero for the decibel scale in air is the threshold of human sensitivity, that level has no relevance underwater, so scientists used a baseline of 1 µPa for decibels in water. Second, water is eight hundred times denser than air and sound travels faster through it; both are factors that combine to create lower intensity for the same pressure. The upshot is that we have to subtract 61.5 dB from any underwater reading in decibels to get its air-based equivalent reading—so, for example, an underwater sound level of 140 dB is equivalent to an air-based reading of 79.5 dB.

Of course, describing sound is not just about how loud it is: we also have ways of measuring what we call pitch, or how high or low a sound is.

Frequency and wavelength

The second core concept to describe a soundwave is frequency, usually defined as the number of waves produced by a source each second and measured using the unit hertz or Hz (one event per second). It is common to see kilohertz (kHz), megahertz (MHz) and gigahertz (GHz) used for different sorts of wave with very high rates of vibration.

Inversely related to frequency is our third core concept, wavelength, a commonly used measure which is simply the distance between a point on one wave and the same point on the next wave. In any given medium, soundwaves travel at a fixed speed, so greater frequency must mean shorter wavelength, because more of them are passing by in a second.

We experience frequency as the pitch of a sound. Technically, frequency is a measurable quantity while pitch is a subjective perception, but for our purposes we can ignore this distinction because the higher the frequency, the higher the tone or pitch you will perceive. An exact doubling of frequency is an octave, which we perceive as the same note but higher up. There is no parallel to this perception in any other form of vibration: we certainly don't perceive ultraviolet light as in any way related to infrared, even though its frequency may be double.

The audible frequency range for humans with perfect hearing runs from 20 Hz or twenty cycles a second to 20 kHz or 20,000 cycles per second. Waves that are outside that range are not detectable by those microscopic hairs in our ears. As we get older, the shorter hairs that measure higher frequencies tend to fail first, which means we tend to lose our ability to hear higher-pitched sounds. Somewhat arrogantly, we humans term any sound below 20 Hz as infrasound, and anything above 20 kHz as ultrasound—but, as we'll see later, for many other animals on this planet, those frequencies are mainstream.

Like all waves, soundwaves can be diffracted: they will bend around obstacles that are smaller than their wavelength, so lower frequencies with their longer wavelengths do this better than very high frequencies. If you're playing hide-and-seek and standing behind a tree, you will still be able to hear the "Ready or not, here I come!" of the seeker on the other side of your hiding place, though you might not hear them hiss or tut.

When sound meets a surface like a wall, it may do any or all of three things. First, some or even all of it may bounce back: the harder and flatter a surface, the more likely this is to happen. Second, some sound may be absorbed by the surface it meets. Energy never vanishes, so in this case, the sound energy turns into heat—though soundwaves carry very little energy, so cranking up the stereo is not a brilliant strategy for heating your house. Soft surfaces like carpets, curtains, drapes, plants and soft furnishings all absorb sound well, making spaces quieter. Third, some of the sound may transmit through the surface to the other side. I'm sure you have been in rooms where you can hear every word the people next door are saying, or stayed in a hotel where corridor conversations annoyingly wake you at night.

The human voice ranges from around 80 to 180 Hz for a typical adult male, and around 165 to 255 Hz for a typical adult female. These are soundwaves ranging from 4.25 meters to 1.33 meters long. Many rooms with hard, parallel walls will have at least one dimension somewhere in that size range, so it's very possible that someone will have a voice at exactly the wavelength that matches one of the room's dimensions, setting up an unpleasant booming resonance as the soundwaves bounce back and forth from wall to wall. You may well have experienced this, so it's useful to know that it can be tempered by placing something soft or irregular to absorb or break up the reflections. A bookcase or some plants can do this quite well.

High and low frequencies also behave differently with solid materials. Stand outside a loud rock concert and you won't hear the hi-hat or cymbals the drummer is hitting, or the high notes on the guitar or keyboards; the bass drum and bass guitar, however, will pound their way through bricks and mortar (and the audience) to reach your ears with ease. This happens for two reasons. First, high frequencies lose energy faster than low ones. Every wave is an increase and then decrease of pressure on the particles in the medium, and each time this happens, it costs energy. Imagine doing sit-ups: the more reps per minute you do, the more energy you expend. Low frequencies keep more energy because they have fewer waves per second, so when they reach an obstacle, they hit it harder and make it vibrate more strongly. Second, large objects like walls tend to have low resonant frequencies, which may well sync with the low-frequency sound, so that deep bass may make walls resonate strongly, amplifying the signal rather than blocking it.

For the same reason, we find it hard to localize low frequencies: they go right through our heads, defeating the clever mechanism we use to identify sound sources, which is all about the timing difference of the sound reaching each ear. That's why it doesn't matter where you position a subwoofer in a room: the bass will fill the room with no discernible source as far as your brain is concerned.

Timbre

How can you tell the difference between a guitar and a trumpet, or even a clarinet and an oboe, playing the same note at the same volume? The answer lies in timbre, or the feel of the sound. Timbre is a sound's character, mainly resulting from its envelope and harmonics, combined with qualities like vibrato and what is known as transients, the short burst of energy you hear at the beginning of sound. These factors combine to make each sound identifiably different.

Envelope describes how a sound changes over time. It consists of four factors: attack, decay, sustain and release, which is often abbreviated as ADSR, which any user of synthesizers will be familiar with. **Attack** is the time it takes for a sound to go from inaudible to its loudest peak. **Decay** is the time it takes for a sound to go from that peak loudness back down to the sustain level. **Sustain** is the loudness of the sound's post-peak level, lasting until the note stops sounding. **Release** is the time it takes for the sound to decay all the way to silence after it stops playing.

In music, both the instrument and the way it's played will determine the envelope we hear.

Almost all sounds have *harmonics*, which comprise a sound's fundamental frequency and a number of other frequencies known as its overtones. The overtones of most musical sounds occur in what's known as the harmonic series. Different instruments produce different overtones, which again determine the familiar character of an acoustic guitar or a trombone.

This brings us back to the choir on the cliff edge that began this chapter, and to a key idea underpinning that experience. Some years after we had sung together in Denmark, I interviewed David Hykes for a podcast series I made, and asked him if harmonics were universal. His answer was unequivocal:

> Yes. There are nineteen constants in the universe and if there were to be a competition to vote for the twentieth constant, I would say harmonic presence certainly would qualify. It is in mathematical terms called the continuum, one of the larger infinite series... [it] always makes me smile when I hear a mathematician say that it's the infinite series of the counting numbers and the fractions—which are, of course in musical terms, not fractions at all, but the intervals that we use, the relationship of one harmonic number to another: two over one for the octave, three over two for the fifth and so on. So this infinite range of whole number relationships does occur absolutely everywhere, and is the DNA of absolutely all musical sound in every octave.

Harmonics are a fascinating aspect of sound that merge music with mathematics and open up realms of philosophical, scientific and spiritual investigation that have occupied both influential thinkers and crackpots for many centuries.

Most sounds, including those made by instruments and human speech, are compounds of many tones that your brain synthesizes into a single experience. Usually, the tone you are most conscious of is the lowest, known as the fundamental. Sometimes we may even imagine the fundamental when it's not really there, as people did with old-fashioned telephone calls, where their brains added imaginary bass to the narrow bandwidth they were actually hearing.

Above the fundamental are overtones, which may be harmonics or partials. Harmonics are a series of tones that sound consonant, or tuneful, to our ears, while partials exist in between the harmonics and may well be dissonant. To an extent, what sounds tuneful or dissonant depends on what you are used to listening to: the Western musical tradition has a different set of rules for what sounds pleasant to other musical traditions. Nevertheless, there is always essential mathematics relating to harmonics.

In seventh-century-BCE Greece, Pythagoras is said to have invented an instrument called a monochord, a single string stretched taut on a wooden frame. He discovered that stopping the string halfway along its length raised the tone by an octave—so that applying a fraction of ½ doubled the frequency, or 2f. Stopping at ⅓ the length creates 3f, which in musical terms is a perfect fifth up from the previous tone; ¼ yields 4f, a perfect fourth up from the previous tone and also a further octave up from the fundamental; ⅕ raises the tone again by a major third. These prime intervals of a third, fourth and fifth are used in much of musical composition.

This harmonic series, also known as just intonation, continues upward, with the gaps becoming smaller. A beautiful graphical representation of part of the series is below, showing the even-numbered string harmonics from the second to the sixty-fourth. This emphasizes how sound, music and art are intimately related with mathematics—the epiphany that is the source of the Pythagorean credo that "all things known have number."

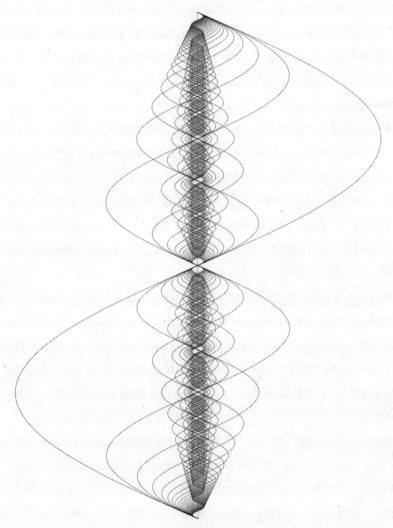

First 32 harmonics of a string.

Like many other philosophers, Pythagoras believed that harmony (he called it *harmonia*) is the purest natural state of the universe and of humanity, the force that creates order from chaos, and the basis of beauty in any form, visual or auditory. To the Pythagoreans, mathematics had four branches: arithmetic and music (both concerned with quantity), and geometry and astronomy (both concerned with magnitude); this grouping was maintained all the way through to the Middle Ages, when the Roman philosopher Boethius named it "the *quadrivium*" and it remained the basis of Master of Arts degrees in medieval universities.

Having discovered the congruence between number and music through harmonics, Pythagoras applied the same rules to the then observable "*kosmos*" (he is thought to have been the first to coin this word for the universe), assigning harmonic intervals to the planets and stars and asserting that the music of the spheres was real, and only inaudible to us because of our imperfections; to the Pythagoreans, human-made music is pleasing because, although a pale echo, it gives us glimpses of what Boethius later dubbed *musica mundana*, the mathematical harmony of the cosmos. More broadly, the Pythagorean belief that mathematics underpins all of reality is a creed shared by most modern physicists and cosmologists, as the universe continues to reveal its wonders in successive elegant equations.

Musically, the Western world (and, thanks to global communication, most modern music everywhere) has diverged from the natural harmonic series since the invention of the well-tempered scale in the eighteenth century, and perhaps in the process it has lost something of its connection with the underlying structure of nature. Some ethnic musical styles are still based on scales using natural harmonics—for example, Indian raga music, Persian *dastgāh* music, Arabic *maqam* music, Chinese *guqin* music, Scandinavian fiddle music and many *a capella* singing styles (the human voice doesn't have frets!).

Soundscapes

It is rare that we experience sound in the real world as a single source. At any one time there are usually myriad soundwaves, like thousands of exploding fireworks overlapping with each other in space and time all around us, and often reflecting off objects in their path. Most of the time, our ears and brain are processing an enormous variety of different types of soundwaves of different amplitudes and frequencies. We have a word to describe the totality of all the soundwaves in any given place at a given time: the soundscape. The auditory equivalent of a visual landscape, the word was popularized by R. Murray Schafer in his classic 1977 book *The Soundscape: Our Sonic Environment and the Tuning of the World*. What Schafer realized was that the totality of sound around any listening being is a powerful experience that's unique to that moment in that place for that listener. In his book *Sonic Experience: A Guide to Everyday Sounds*, Jean-François Augoyard writes, "The concept of 'soundscape' is concerned with the quality of listening [and] what is perceptible as an aesthetic unit in a sound milieu." In other words, a soundscape is a triangular relationship between the listener, the acoustics of the space and the sounds in the space. The International Organization for Standardization (ISO) defines a soundscape as the "acoustic environment as perceived or experienced and/or understood by a person or people, in context."

Soundscape can also be used to describe the characteristic sounds of a specific place. Many natural soundscapes are disappearing as habitats are eliminated or species go extinct. Even urban soundscapes vanish: in London, some of the classic sounds my parents would have known so well no longer exist—ringing bells or two-tone sirens on police cars; tugboat whistles on the Thames; rag-and-bone men calling "Any old iron?" from their horse-drawn carts; the slamming of old-fashioned train carriage doors. Even iconic London sounds I grew up with and took for granted, like "Mind the gap" on the tube or the gravelly diesel engines of black cabs will not last forever. The British Library Sound Archive is a collection of over six million sound recordings dating from the dawn of audio recording and includes gems like J. R. R. Tolkien visiting a tobacco shop in 1929. In 2010 it sponsored a project to capture the sounds of Britain, and 350 contributors uploaded over two thousand recordings. But to compare this to the preponderance of visual preservation in photographs is to see how much of the sound that forms our auditory heritage has been lost forever.

So far in this chapter we've looked at the mathematical perspective of our experience of sound, including the range of frequencies we can hear as pitch, or experience as volume and measure in decibels. But as we'll see again and again, everyone listens differently, and soundscapes are always to some degree personal and subjective. Your music might be my noise; your enjoyably loud song in a car might stop me concentrating on driving. So many of the soundscapes we experience are accidental by-products of activity in spaces that were designed with little or no thought given to how they sound, so it's unsurprising that we will be discussing noise, defined as unwanted sound, throughout the book. Most human-made soundscapes are unintentional, undesigned and unpleasant. At their most extreme, we will see our soundscapes are actively endangering our own hearing and causing terrible harm to other living things.

WHAT YOU CAN DO

> Pay careful attention to the soundscapes that surround you as you move through your day. Assess the elements of each, and ask yourself if this is the most beneficial, productive and healthy environment you could be in. You may find some surprising answers.

*

We have only started to encounter some of the myriad ways that the sound impacts how we feel, think and behave. It's time to start our journey through some of the wonders of sound so that we can better understand its effects on us, and experience how wonderful the world of sound is if you just start to listen consciously. We'll start with the sound of living things, or biophony.

4

BIOPHONY

Sound is the vocabulary of nature.
Pierre Schaeffer

When the Sun comes up over the Kwando river in Namibia, as the light spills over the land, so does the sound. At first an irregular peeping, then a kind of deep clicking or rattling, then whoops and cries, with increasing urgency, different rhythms, calls rising and falling in volume. Within minutes, the sound thickens, a filigree of light, distant calls punctuated by the sudden punch of a caw—then glugging, rasping, screeching, roaring, grumbling, as thousands of animals come into awareness and begin to locate themselves and each other. To listen to the world wake up like this and move from silence to a rich symphony of noises is to rediscover how tightly bound life is to sound.

The world is alive with the sounds of living things, though tragically less so every day as the remorseless impact of humanity on biodiversity plays out in every habitat on the planet. Hearing has been called "the universal sense" by Seth Horowitz in his wonderful book of the same name. But why is it so universal?

Perhaps the main reason animal sound is so ubiquitous and so diverse is that vibration is always around us. Light comes and goes, depending on where you are and when. At the poles, the longest day eventually gives way to an interminable night, while at the equator we have the startlingly brisk handovers from night to day and back again at exactly six o'clock, and half of every day is lived in darkness. Even in daytime, dense jungle and rainforest shade around eight percent of the world's land surface, creating habitats that are dimly lit even on sunny days, while many caves and all the ocean deeps are perpetually pitch black.

The sense of touch is by definition limited to any organism's immediate vicinity, while chemical sensory tools like smell and taste are also relatively short range in most cases, although there are exceptions: African elephants have been reputed to scent water almost twenty kilometers away.

By contrast, vibration can exist in any context other than a vacuum and can be sensed over great distances, often supplying rich and detailed information about an animal's environment. One thing Charles Darwin has taught us is that if there is a source of information, then life will evolve to make use of it. So perhaps it should come as no surprise that of the seventy thousand-odd vertebrate species on Earth, including mammals, reptiles, birds and fish, there is not one that lacks a functional sense of hearing and vibration—while there are many with limited, or even completely absent, senses of sight, smell, touch or taste.

Complex environments like rainforests come alive sonically at night, when light is of no help. The animal orchestra of a rainforest is huge and diverse, and each player is determined to be heard, even over the deafening sound of the rain which so often falls.

Distinguishing a critical signal (possibly the sound of a predatory jaguar) from irrelevant noise requires acute hearing that's tuned to the key frequencies, neurology that delivers the information quickly and accurately, and conscious, attentive listening skills. Animals don't just hear, they listen.

Sound is clearly integral to vertebrate life. Many animals have learned to evaluate the volume, direction and meaning of the sounds around them; as we have noted, hearing is a spherical sense, which makes it the only way of sensing what's behind you—and hearing delivers its information to the central nervous system far faster than sight. As the primary warning sense and a key way of hunting for food, hearing is a matter of life or death for many animals. And sensitivity to vibration starts far below the order of vertebrates, at the smallest unit of life.

The smallest sound

At some point, at least 2.7 billion years ago, a new kind of cell formed—one with a nucleus enclosed within a membrane. Such cells went on to form you, me and the other animals with backbones and a skeleton—that is mammals, birds, reptiles, amphibians and fish—as well as the invertebrates, including insects and crustaceans, alongside one million species of plant and fungi. There are an estimated 8.7 million species whose cells are structured like this. Only about one percent of these species, known as eukaryotes, are vertebrates. To date, we've identified around one million of the eukaryotes, leaving almost eight million yet to be named. Even this vast panoply of life is dwarfed by the likely number of species whose cells do not have a nucleus in a membrane, known as prokaryotes. Some estimates suggest that there are as many as one trillion species on Earth, the vast majority of them microbial prokaryotes. It's time to enter the world of single cells, which, as we will discover, is a surprisingly noisy place.

Sonocytology, the study of the sounds that cells make, is a recent discipline. It was founded in 2002 by Professor James K. Gimzewski, Distinguished Professor of Chemistry at the University of California, Los Angeles and Director of the Nano & Pico Characterization Core Facility of the California NanoSystems Institute. Gimzewski has specialized in nanoscale science for over forty years, pioneering the imaging of molecules using scanning tunneling microscopy and latterly the detection of the sounds cells make, using an atomic force microscope (AFM).

An atomic force microscope can image objects smaller than one nanometer (nm). To give you an idea of how extraordinary that is, a human hair is around 80,000 to 100,000 nm wide. No optical microscope can see things as small as a nanometer because of something known as the "diffraction barrier," which restricts a microscope's ability to distinguish between two objects separated by a distance smaller than about half the wavelength of light. The wavelength of light is 380 to 750 nm, so the smallest object we can ever see with light is around 200 nm in size. The AFM achieves two hundred times this resolution by using touch rather than sight. Its tip is less than 10 nm wide; like a minuscule record-player stylus, it deflects as it moves across the surface of the tiny object being scanned. The tip's movements are detected by a laser beam, and the accumulated data creates a detailed map of the surface of the nano-object. In order to eliminate external vibrations, Gimzewski's AFM is housed in a special darkened noise-free room, inside a thermally, acoustically and electrically isolated chamber that's lined with foil and mounted on a vibration-free platform suspended in air.

When Gimzewski and his graduate student Andrew Pelling held the tip of their AFM against the surface of a yeast cell, the device recorded that the wall of the cell was rising and falling by 3 nm around a thousand times a second. They used a free sound-conversion software utility to convert this vertical deflection data into an eerie audible tone which, at 1,000 Hz, is roughly an octave above middle C.

Worried that this sound was an artifact of the equipment they were using, they soaked yeast cells in rubbing alcohol, which breaks the cells' membranes and kills them. The cells "screamed" at a much higher pitch, and then went silent as they died, proving that the sound was indeed cell-generated and, as you can imagine, creating some lurid headlines in the science sections of the world's press.

Next, the team varied the conditions of the experiment and the cell types they examined, and came to the extraordinary conclusion that each cell produces a unique sound which comes from inside the cell itself. Their theory (as yet untested) is that the sound arises from the movement within the cell of motor proteins that carry out a range of tasks, such as transporting tiny structures known as organelles within the cell. The movements of these motor proteins occur at a thousand steps a second—exactly the frequency detected by the AFM.

Whatever the mechanism, the discovery that all cells have unique voices is both rather wonderful and potentially very useful. Cancerous cells have a softer cell membrane, so they may have a distinctive sound with a lower tone; this may also be true of other diseases. Sonifying one cell could be enough to diagnose a disease in the future, if and when scientists create a reference library of cell sounds.

Sadly, progress since the publication of the original Gimzewski and Pelling paper in 2004 has been slow, perhaps yet another example of sound being undervalued in an ocular world. But there are an increasing number of potentially ground-breaking applications for the sonification of cells.

A team at Manchester University led by biologists Richard Snook and Peter Gardner has been testing another way of translating cell qualities into sound to spot cancer: bombarding human prostate cells with infrared light. As the cells rapidly heat up and cool down, they expand and contract, vibrating the air molecules around them. Statistical analysis of the thousands of simultaneous notes that arise allows the team to spot cancer cells. "The difference between a healthy cell and a cancer cell is like listening to two very large orchestras playing their instruments all at the same time," Gardner says. "But in the cancerous orchestra, the tuba is horribly out of tune." The team's aim is to greatly reduce the number of painful prostate biopsies, seventy-five percent of which return negative results.

A team at the University of Missouri-Columbia showed in 2006 that they could detect early signs of metastases from melanoma by listening to cells. They vibrated a sample of white blood cells with rapid-fire sequences of brief but intense blue-laser pulses, each lasting just five billionths of a second. The dark, microscopic granules of melanin contained in the cancer cells absorb the energy bursts, rapidly expanding and shrinking as they heat and cool. These sudden changes generate cracks which propagate in the solution like tiny tsunamis. This method can spot as few as ten cancer cells in a sample, and it takes just thirty minutes.

More recently, a team at the University of California, San Diego, have advanced the field by creating an optical fiber a hundred times thinner than a human hair, coated with a thin polymer layer studded with gold nanoparticles. They dip this glamorous-sounding probe into a solution containing cells and shine a light down the fiber. As tiny vibrations from nearby cells press the gold nanoparticles into the fiber, the team measure the resulting minuscule changes in the intensity of light. They estimate their tool is ten times as sensitive as an AFM, and is able to detect forces of less than 160 femtonewtons (each one of these tiny measures being one ten-trillionth of the force an apple exerts sitting on the ground) and sounds of less than -30 dB, effectively one thousand times lower than anything the human ear can detect. "We're not just able to pick up these small forces and sounds; we can quantify them using this device," claimed the team's Donald Sirbuly. The aim is to create what would effectively be a nanostethoscope, with the objective of detecting cancerous cells quickly and accurately. Perhaps we will soon be able to hear our own cells singing their unique songs.

A recent paper reveals that some bacteria create sound by flailing their tail-like structures called flagella. The tiny movements are, of course, far too small for us to hear acoustically—a problem that Farbod Alijani's team at Delft University of Technology brilliantly solved by using ultrathin graphene drums. Graphene is a layer of carbon that's just one atom thick and is very sensitive to vibration. The team found that a single E. coli bacterium adhering to a graphene drum thumped the surface in random patterns with an amplitude of up to 60 nm. "To understand how tiny these flagella beats on graphene are, it's worth saying that they are at least ten billion times smaller than a boxer's punch when reaching a punch bag. Yet, these nanoscale beats can be converted to sound and listened to," Alijani says [3].

The team found that when antibiotics worked to kill the bacteria, the drumbeats faded over two hours and then stopped altogether, while if a bacterium was resistant to the antibiotic it drummed on. Antibiotic resistance is a major threat to human health, arising from decades of indiscriminate over-prescription, and this could be a quick and accurate way to test bacteria for it, using just a single cell.

So, cells make sound—but do they perceive or respond to it? Very little research has been carried out on this, but a 1998 paper from Tokai University found that soundwaves of 6 to 10, 18 to 22 and 28 to 38 kHz promoted colony formation of *Bacillus carboniphilus*, and, even more fascinatingly, that cells of *Bacillus subtilis* were emanating sound at frequencies of 8 to 43 kHz with broad peaks at approximately 8.5, 19, 29 and 37 kHz—very similar frequencies to those stimulating the growth of the other bacteria. The intriguing conclusion was that cells may be signaling among themselves with these sounds to regulate or encourage growth.

In 2021 scientists from the University of Auckland found that sound at 100 Hz and 10 kHz caused brewer's yeast cells to grow twenty-three percent faster, and to produce more aroma compounds associated with citrus notes. The authors suggested that sound could be used in brewing to affect both the timing and the taste of the final product.

The response of cells to sound may be under-researched by mainstream science, but there is a large spectrum of people who passionately believe that there is a strong connection between sound and health, with the mechanism being cellular entrainment. This includes some scientists, such as the University of Toronto's Professor Emeritus of Music Lee Bartel, whose TEDxCollingwood talk (albeit issued with a warning from TED that its claims are far from substantiated) presents some evidence of beneficial effects of a tone of 40 Hz on sufferers of Alzheimer's and fibromyalgia, and suggests other potential uses for this sound.

The community of passionate believers in the links between sound and cellular health also encompasses the many who practice or receive sound therapy or vibrational healing from chanting, singing bowls, gongs, tuning forks or other methods. Most of these practices are based on the established fact that cells vibrate, and the hypothesis that health requires these vibrations to be in some sort of harmony; disease is defined as disharmony which can be ameliorated or even cured by using an external sound source to entrain the out-of-tune cells back into their healthy state. As I wrote earlier, it wouldn't surprise me at all if in the future mainstream medicine discovered that health and harmony are indeed related, and refined these ancient healing methods to create a scientifically validated body of sound-healing techniques. Time will tell. In the meantime, what is certainly true is that as small as we can measure, there is sound. It seems to exist wherever we look, even in the plant kingdom.

WHAT WE CAN DO

Sonocytology has so much potential that we must encourage and fund research in this exciting field, because there are enticing benefits to be had in the diagnosis of human and plant disease, in food production and in scientific understanding of well-being.

Green noise

Below is a diagram of the Earth's biomass. By far the largest kingdom, at over eighty percent of the biomass, is plants, then bacteria, then fungi, then archaea and protista, two different sorts of single-celled organism. Animals are the smallest. And humans? We're a tiny subsection of that smallest category.

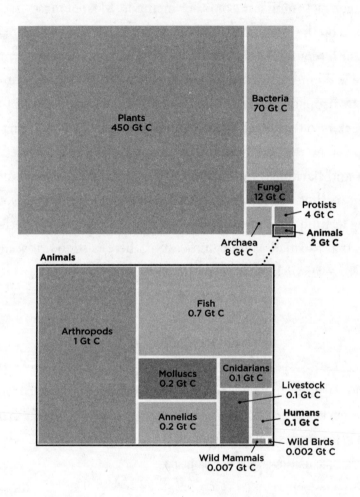

Sound appears to be universally important across all these domains. There is evidence that plants respond to vibration—and some even create sound.

Thousands of plants have evolved a response called buzz pollination, in which they release their pollen when they are vibrated at the exact frequency of bee wingbeats. Some plants have even learned to make their nectar taste sweeter when they sense such vibrations, or at times of peak pollination activity by insects.

Buzz pollination is a very understandable evolutionary adaptation that makes life easier for both parties in the transaction. But how do we explain plant roots bending to grow toward a particular sound? In laboratory experiments, scientists have played continuous tones coming from one side of the roots and observed them bending toward the sound. The response is most vigorous at 200 to 300 Hz. Even more fascinating: the roots themselves are emitting sound in the form of detectable clicks as they grow.

Most plants create some sound simply as part of their living process, or as they respond to their environment. We perceive wind largely second-hand, as trees creak, leaves shimmer or grasses susurrate, while the ardent new wave of nature sound recordists immersing their hydrophones in ponds are hearing many plants bubbling as they release gas. These are all consequential sounds, the by-products of another activity, not intended as communication.

However, it now seems that at least some plants do create sound intentionally, and possibly with the aim of communicating. Itzhak Khait and his team at Tel Aviv University have measured tomato and tobacco plants broadcasting ultrasonic sounds at 20 to 100 kHz when stressed by a lack of water, or when their stems are cut. The sounds were picked up by microphones placed ten centimeters from the plants, and the team estimate that ultrasound-sensitive insects and mammals could hear them from up to five meters distant. On average, drought-stressed tomato plants made thirty-five sounds an hour, while tobacco plants made eleven. When plant stems were cut, tomato plants made an average of twenty-five sounds in the following hour, and tobacco plants fifteen. Unstressed plants produced fewer than one sound per hour, on average.

In the past, careful recording has discovered an ultrasonic sound coming from drought-stressed trees, created by cavitation—air bubbles forming and collapsing inside xylem tubes—but this was not audible outside of any tree. It seems likely the tomato and tobacco plants are using the same mechanism, but the researchers speculate that radiating the sound may act as a warning and change the behavior of moths seeking a well-hydrated plant on which to lay their eggs, for example. The function of a plant screaming ultrasonically when it is cut is harder to pin down without getting seriously anthropomorphic.

Sound production by plants is barely researched and little understood. Given that cells vibrate and that the ground can carry sound and other vibrations very well and over long distances, it would be surprising if plants didn't take advantage of this sensory pathway, along with chemical and light sensitivity. Sound is fast and it costs very little energy to make it—far less than generating chemical signals, for instance.

Despite the recent surge in interest, we still have so much to learn about plant senses and communication. Because they don't move in our timeframe, it's so easy to dismiss plants as almost inanimate. But time-lapse photography can bring plants to life in the most visceral way; they reveal themselves as active predators, prey or cunning strategists, rather than static decoration. Recent science has revealed startling evidence about the strength and depth of communication between trees through the mycelium networks connecting them in the soil, and the complexity of the electrical signals moving through these fungal networks, while a recent study found that some species of fungi grew faster if certain sounds were played to them. More and more, it seems that these electrical impulses are akin to a language with words.

Really, we should not be surprised if plants are using sound to communicate. It seems that almost nowhere we look in the natural world is there silence.

Now it's time for us to turn to the smallest organism we humans can hear with the naked ear: the insect.

What's the buzz?

Though they are a small category in comparison to plants, fungi and microbes, insects are still the largest single subsection of animal life. The famous naturalist E. O. Wilson estimated that the number of insects alive on Earth is ten quintillion (10,000,000,000,000,000,000) which is roughly the number of stars in one hundred million galaxies; insects outnumber us by well over a hundred million to one. There are estimated to be as many as thirty million species of them, most of them as yet unnamed by scientists.

Insects are hugely versatile and resilient. They exist everywhere, from sand wasps and ants in the hottest deserts and hot-spring midges that survive temperatures of over 80°C, to Antarctic midges, snow fleas and snow crickets. Some insects even spend part of their lives underwater.

Most insects have a major challenge to overcome when it comes to making sound: their size. If you're very small, making loud sound is only possible at high or even ultrasonic frequencies. For this reason, insects occupy a space in the great animal orchestra that starts where the highest-pitched vertebrates leave off. It is now known that many insects overcome the challenge of being small and yet needing to be heard at a distance by using substrate-borne vibration, often causing plants to carry their sounds over distances up to several meters.

There are six principal ways insects make sound. *Stridulation* is rubbing body parts together, possibly mandibles, legs, or some sort of evolved specialist file-and-scraper assembly. Crickets are the most common example. *Percussion* is banging on something resonant, usually the substrate, possibly in a group with synchrony—some ants and termites make a surprisingly powerful sound this way. *Tymbalation* involves flexing a rigid body part back and forth like a tin lid—this is the party piece of cicadas, and also used by some moths to jam bat sonar. *Tremulation* is the rhythmic movement of the whole insect, or possibly flight muscles, often passed via the substrate. *Flight* is the familiar buzzing of bees or whine of a mosquito. And, finally, *forced air* involves pushing air through spiracles or the alimentary canal to create a sound. Some cockroaches use this sound to hiss, and the death's-head sphinx moth employs it to make a sound like a dog's squeaking toy. You can enjoy many examples of insect sound on the Songs of Insects website [4].

Let's start with the champion insect sound-maker: the cicada, not to be confused with the gently chirruping cricket. Crickets (and some ants) create their noise by using a kind of natural guiro, drawing parts of their body called scrapers across other parts called files to create a short ratchet sound, usually delivered in a repeating rhythmic manner. The tempo of cricket stridulations is directly related to the temperature. Crickets are cold-blooded, and chirp faster as they get warmer in a predictable way, so if you know the species of cricket, you can tell the exact temperature by counting the chirps in a fifteen-second period and adding a number that's specific to that species. One theory for why humans have certain rhythmic sensibilities is that they come from listening to these kinds of insect sounds: crickets and other orthopterans such as grasshoppers, locusts and katydids will often congregate and chirp in large groups, with the sound coalescing into a rhythmic synchrony that can be almost mesmeric.

While crickets make quite a gentle sound, cicadas are another thing entirely. They have a mysterious mating cycle in which they can spend as long as seventeen years underground before emerging and looking for a mate. On reaching the air, they start seeking their mate by flexing a rigid membrane called a tymbal that clicks like a metal tin lid being pushed in and out—except they do this more than three hundred times a second, using a powerful muscle and amplifying the sound with a resonating chamber like the body of a guitar. The result is often over 100 dB of whine or buzz from a single cicada, a sound that can be audible almost a kilometer away. But you rarely encounter a single cicada, because the male cicadas tend to hang out in large gangs—and when one starts to sing, the rest quickly join in, until you have a high-pitched roar that can make conversation impossible. Cicadas have rudimentary eardrums located on their sides. In order not to deafen themselves with their own song, male cicadas have a reflex where a muscle contracts to fold their ears shut as soon as they start to sing. After their years spent underground in the nymph stage, the males' urgent song to find a mate adopts a rather tragic air when you discover that immediately after mating, they die.

Many orthopterans can hear very well. Crickets have their eardrums on their front legs, and recent research by entomologist Professor Fernando Montealegre-Zapata at the University of Lincoln has discovered that they, and probably other orthopterans, have a structure that, like the bones in the human ear, amplifies sonic vibrations and then transduces them from air to liquid, increasing the frequency range they can hear as well as their sensitivity. He estimates that these insects have been singing and hearing since the Triassic era, 250 million years ago. Crickets typically make and receive sound in the 2–8 kHz range, while katydids can hear all the way up to 150 kHz—the same upper limit as dolphins, and around three octaves higher than humans can hear. This is probably so that they can detect the high-frequency echolocation sounds made by predatory bats.

Some Australian katydids have found a more sinister use for their sound-making skill: they have developed what's known as "aggressive mimicry" to prey on those desperate cicada males. The cunning katydids imitate the species-specific replies of sexually receptive female cicadas. They do this not only acoustically with clicks made by their thick forewings, or tegmina, but also visually, with synchronized body jerks. Remarkably, the katydids respond appropriately to a variety of complex, species-specific songs, including the songs of cicada species they have never encountered before. The unfortunate male cicada that replies meets an unfulfilled end instead of the anticipated culmination of his existence.

Singing is not without peril for other insects. Experiments have shown that parasitoid female tachinid flies listen for the calling song of field crickets from dusk to dawn. The flies are phonotaxic, which means they move in response to sound. In this case, they use the sound to navigate to an unfortunate cricket and deposit their larvae on it; the larvae burrow through the cricket's exoskeleton, grow for a week and then burst out, killing the host.

Numerous insect predators—from frogs and lizards, to cats, birds and flies—use insect sound to hunt. A 1978 experiment in Ocho Rios, Jamaica, established that the little blue heron uses cricket song to stalk, catch and eat the singers. In Florida, scientists have discovered that considerably fewer male crickets sing in the autumn, when the parasitic flies are particularly abundant. They also found that female crickets are reluctant to approach singing males at this time of year.

Peril is part of insect life, as it is for so many species. Numerous insects have evolved defensive sounds that are designed to frighten, intimidate or confuse predators, or possibly to warn others in their group of the danger. The synchronized singing of massed crickets is one strategy, also used by croaking frogs: in the dark, a predator can't identify a single entity to attack due to the Phil Spector-like wall of sound that greets it as the group maintains a kind of force field of bewilderment [5]. The moment the synchrony breaks, everyone is potentially dinner.

Individual insects have their own acoustic defenses too. The Death's-head hawkmoth and some Madagascan cockroaches hiss; tiger moths click; some caterpillars force a loud whistling sound from their alimentary canal to scare birds, which may be the only example of aggressive farting in all of nature.

Some insects even make sounds of triumph in battle. In a filmed conversation with Sir David Attenborough at the British Library, Chris Watson recalls recording the hissing stridulation made by a column of South African Matabele ants as they returned carrying many dismembered termites from a successful raid. The ants attack a termite nest in formation and in complete silence, but when victory is achieved they celebrate on the way home with their hissing song, like Vikings returning with the spoils after sacking yet another monastery [6].

Of the thirty major insect orders, nine are known to include insects that can hear. Associate Professor of Molecular Biology Martin Göpfert at the University of Göttingen in Germany studies hearing in the fruit fly. He estimates that hearing has evolved independently more than twenty times in insects. This may account for the astounding variety of locations for the hearing organs. As we've seen with cicadas and orthopterans, ears can be on legs and abdomens—the bladder grasshopper beats the cicada by having not one but six pairs of ears along its abdomen—but they can occur almost anywhere: mosquitoes, bees and fruit flies have a group of vibration-sensitive sensory cells called the Johnston's organ attached to their antennae; hawkmoths have their ears in their mouth parts; while a praying mantis has its one ear in the middle of its chest. Several insects have developed the same mechanism used by crustaceans, with mechanoreceptive hairs protruding at different angles from their bodies or legs, allowing them to perceive the strength and directionality of air vibrations.

In every case, insects have evolved to hear what they need to. Mosquito ears have a range of about a meter and are tuned to detect the tiny whine of another mosquito flying, which is how they find a mate; the bladder grasshopper with its twelve ears can hear sounds from up to a kilometer away, including the wingbeats of predatory birds. Mantis and moth ears appear to have evolved the ability to hear bats' ultrasonic echolocation frequencies once bats appeared around fifty to sixty million years ago—before that, scientists are divided on whether these insects either had no ears at all, or had developed ears for conspecific communication only, and then extended their range into the ultrasonic as bats became their primary threat. Having evolved a hearing ability that keeps them safe, insects now also use it to make sound for any or all of the usual reasons: territory, mating, food and danger. In 2013 a team at Strathclyde University in Glasgow found that the greater wax moth can sense frequencies of up to 300 kHz, making it the current world champion in the highest hearing stakes, almost a full octave above porpoises, dolphins and katydids.

Often confused with insects, arachnids like scorpions and spiders are a different order entirely and their relationship with sound is little researched. Although no arachnids have yet been found to possess ears, experiments have established that they can both detect and interpret sound. Scorpions and spiders are thought to receive vibration signals through hairs on their legs. In one experiment, jumping spiders reacted to the 80-Hz buzz of a mud dauber wasp (their primary predator) with a neural spike and frozen behavior. They did not react to other sounds in this way, so there seems to be some degree of processing and interpretation occurring. Since spiders rely on detecting the faintest vibrations in their webs to alert them to possible captured prey, it's not surprising that they are also sensitive to airborne vibration. Scientists have established that the charmingly named ogre-faced spider can sense sound frequencies from 100 Hz to 10 kHz, which is comparable to human hearing.

The vast majority of arachnids make no sound, but there are two rare exceptions. A male wolf spider uses its pedipalps (jaw-like structures) to generate a low purring vibration that it propagates through dead leaves, amplifying it to a level that even humans can hear, in order to attract females [7]. A tarantula has a different goal when it produces a hissing sound by rubbing its pedipalps and forelegs together: this is a warning sound to scare off potential threats.

To most of us, the most familiar sound from the insect kingdom is the buzz, whine or whir from wingbeats in flight. A summer meadow would be incomplete without the buzzing of bees, while a house fly trapped in a room is an all too familiar distraction when working, and the high-pitched whine of a mosquito by one's ear is guaranteed to wreck any chance of sleep. Bees and flies beat their wings in the region of two hundred times a second (200 Hz); large moths can be as low as 50 Hz; while some small flies are as rapid as 2,000 Hz.

Not all insects make sound that's audible to humans when they fly; in Scotland in the summer the famous biting midges are so tiny that they are not only invisible, but impossible to hear. However, they certainly can hear one another with their Johnston's organ picking up vibrations from their antennae, which are tuned to detect the frequency of other midges' wingbeats—roughly 1,000 Hz. Every year in Orkney we also experience a short bloom of crane flies (daddy-longlegs), which are many times larger than midges, and yet are equally inaudible in flight unless they happen to be very close to your ear.

Some bees make extra use of their buzzing. Cavity-nesting bees live in near darkness, so it's hard to see the famous waggle dance, where a returning worker gives the rest of the hive directions to a food source. To make its map clear even in the half light, the dancing bee adds an extra level of communication to its movements by varying its buzzing frequency and strengths. The BBC made a fascinating short film on bee dancing [8].

Fascinatingly, it has been established that the Johnston's organs of the watching bees detect not only this changing sound coming from the dancing bee but also the modulating electrical fields it emits. It's not just bees that do this: you and I could be described as electric fields that hold all our atoms together, and electrical impulses are firing through us constantly as we perceive, think and move. I wonder if we will discover in the future that humans, too, have some ability to detect modulating electrical fields in other humans—a whole new level of nonverbal communication that might explain a lot about intuition and irrational feelings like instinctive dislike or falling in love.

Insect sound is endlessly interesting, but in the academic world it is still a rather poor relation to more glamorous orders. One man who's passionate about insects, and particularly grasshoppers, is Ed Baker, a researcher at London's Natural History Museum. Some years ago, he came across a set of cassette tapes recorded in the 1960s and 1970s by a predecessor named David Wragg. The tapes contained field recordings of European grasshoppers and proved to be the tip of an exciting iceberg: in the museum's archives, Ed unearthed some nine hundred tapes in all, with every recording linked to a specimen that still existed in the museum's collection. This has opened a new, richer style of curation called "the extended specimen," where a static (and always dead) example of a creature is embellished not just with its morphology—a detailed description of its form that is equally static and somewhat prosaic—but also its living behavior. This surely is the future of museums in a digital, multimedia age.

Ed set about digitizing the tapes, and a valuable online resource called BioAcoustica was born. It now contains over four thousand sounds that are publicly accessible and searchable via a comprehensive classification system, with links to scientific papers wherever possible. This is something of a nirvana for anyone fascinated by nature sounds, so make sure you are not on a deadline when visiting his website [9]. The team's intention is to create a sort of Shazam for nature sound, where you can input a sample and the system will find recordings that sound similar.

There are other significant databases of nature sound, including the extraordinary private collections of Bernie Krause, Chris Watson and Martyn Stewart, plus the BBC's Sound Effects library [10]—and, towering over them all, Cornell's Macaulay Library, which at the time of writing has over 1.5 million recordings of many thousands of birds, plus increasing numbers of insects, fish, frogs and mammals, all publicly accessible [11].

Now let's turn to the animals that eat an estimated 500 million metric tons of insects a year.

The mastersingers

Birdsong has existed on Earth for almost seventy million years. Tyrannosaurus rex was an early audience for the first singing bird, a goose-like creature called *Vegavis iaai*. Fossil evidence found on Vega Island, Antarctica, shows that this creature had a syrinx, the specialist voice organ that modern birds use to sing with. Although birds evolved from much more ancient dinosaurs than T. Rex, by the time of the late Cretaceous age the two species coexisted—but only the birds made it through the Chicxulub extinction event sixty-six million years ago.

There is no evidence that any dinosaurs had a syrinx or sang; any sound attributed to them is highly speculative. However, scientists can hazard a guess at the noise this early bird might have made with its asymmetrical syrinx—it was probably a honking sound, not unlike modern ducks or geese.

In the many millions of years before any human ears could appreciate their singing, birds evolved an astonishing range of sounds:

the harsh caws of corvids; the simple chirps of sparrows; the soft cooing of doves; the piercing screech of a hawk; the strident *ke-ow* of a peafowl; the glorious melodic songs of the nightingale; and the extraordinary mimicry of the lyrebird. But why? Other animals have the same challenges of attracting a mate, defending territory or warning of danger; some have been around for almost as long as the birds, but don't sing in such an elaborate way. The Macaulay Library has around 2.5 million recordings of birds; you can jump in and sample them for free at any time [12].

One man who's made a lifelong study of non-human singing is David Rothenberg, who straddles the worlds of science, art and philosophy as a professor at the New Jersey Institute of Technology, a jazz musician and an author who has published books such as *Why Birds Sing*, *Bug Music* and *Survival of the Beautiful*. He has also performed or recorded with humans as diverse as Peter Gabriel, Suzanne Vega, Scanner, Elliott Sharp and the Karnataka College of Percussion—as well as with many birds, whales and insects.

I first met David when he was in London for an event hosted by the author and musician David Toop. In his talk, he described playing his clarinet in duets with nightingales, and he played spellbinding slowed-down birdsong, which revealed the shimmering beauty and intricate, quicksilver artistry of the tiny singer's creation. In subsequent conversations, I wanted to understand more about the unique perspective he has developed in his ear-to-ear encounters with various animals.

It starts with our semantics, because the moment we label the sound of another species as music or song, we are creating a projection that may color our attempts to understand what's really going on. David explains: "I think that one view of evolution puts forth the idea that most of the communication in the animal world is more like music than it is like language, in that there are so many animals that perform. They make their sounds and they perform them over and over again, more like a piece of music than a language where you're interacting and responding and seeing specific messages. The performance is key—and this could include insects, which could be our first sources of rhythm: frogs and crickets are making these repeating simple patterns that together form a kind of emergent complexity. And you have many birds that are presenting the same sound patterns repeatedly, like a piece of music with a beginning, a middle and an end, with a real passion, energy and shape to it. And that's why these sounds are called songs in most human languages."

David is convinced that birds have an aesthetic sense (he's not alone: for much of *The Descent of Man*, Darwin saw the singing bird as presaging the development of human aesthetics). Going further, David believes that many animal species appreciate beauty and strive to create it in their sound, movement, as well as their shapes and colors. Of course, beauty is in the eye (or ear) of the beholder, so it may be that naked mole-rats appreciate beauty in one another in a way we don't understand—or perhaps aesthetics is only present is some species. Nevertheless, as David is fond of saying, explaining the song of a nightingale in terms of mating or territory is just not very interesting: it fails to touch the beauty. There is surely something else going on when a humpback whale sings for an entire day nonstop.

Unlike whales, birds have tiny brains, typically the size of a walnut. As with bats and their instant audio processing to create a mental image of their surroundings, it is astounding that a brain this size can create music as complex as the song of a thrush or nightingale. Research in the 1960s on canaries established that birds' brains forge neural connections when they learn new songs—a realization that led to the recognition of human brain plasticity, where before it had been thought that adult brains stopped developing. It's also true that human brains grow when we learn a new song, which may in part account for the fact that musicians have slightly larger brains than non-musicians.

Birds have ears not unlike our own, with outer ears hidden in the feathers on either side of their heads, eardrums (protected in the case of songbirds by a second membrane), ossicles (the three bones in the middle ear) and cochleas. The cochlea is tiny in most birds but dramatically larger in owls, which have such sensitive hearing that they can detect the rustle of a mouse moving in grass many meters away.

However, birds don't have good high-frequency hearing. Although we think of birdsong as quite high-pitched, almost all of it is under 12 kHz, and the vast majority is in the midrange of 1–8 kHz. Most birds have hearing that's particularly sensitive in this midrange and don't hear very high frequencies at all; however, there is evidence that they can perceive infrasound. In 1979 scientists found that homing pigeons are over thirty times as sensitive as humans to frequencies below 10 Hz, and can perceive sounds as low as 0.05 Hz (with sound traveling at 343 meters per second at 20°C, that's a wavelength of almost seven kilometers!). Four out of five birds tested have this same ability, including even domestic chickens, so it seems likely that most birds are infrasound sensitive. This is not for communication: unlike elephants, they can't make such low sounds themselves. The most likely reason for this ability is navigation, especially for migratory or pelagic birds.

As we will see later, our planet is full of sound with frequencies lower than the lowest limit of human perception, or infrasound. We may not be able to perceive it, but birds can hear infrasound such as microbaroms, a class of sonic waves generated by wave action in the oceans, which can travel thousands of kilometers downwind. Waves breaking on shorelines also generate infrasound. If birds can hear this, it may help to explain how they can navigate huge distances over featureless open water in their migrations, like the graylag geese that arrive here in Orkney every winter from Iceland or the largely pelagic and very charming puffins that nest each summer with their lifelong mates in the same spots on the cliffs here, having overwintered as far south as the seas off North Africa. It's fascinating to wonder what these little birds create in their brains from the infrasound they can hear. Is it a mental picture of distant shores, like a sonic GPS? Perhaps one day neuroscientists will tell us; for now, we can only acknowledge yet another aspect of reality in sound that our human senses simply don't perceive.

Birds have a unique piece of equipment: the syrinx (they do also have a larynx, which retains its function as a respiratory valve, but it does not produce sound). This structure allows songbirds to make two sounds at once, spanning the two forks of the trachea as it bifurcates into the lungs. The left fork produces the lower notes, and the right the higher ones. Very few humans have mastered the trick of producing two notes simultaneously with our single larynx, though elsewhere in this book we have encountered some who can achieve this through overtone singing.

It's clear that birds understand rhythm, given that it is a key element in learning their particular song, though their sense of rhythm is perhaps rather different from ours. But do they respond to repeating regular rhythms? David Rothenberg has played with birds for many years so his view is informed by unique experience:

"I would say anyone who spends time with them would say, yes, they do all the time, but it's not generally accepted by science that they do."

Equally, harmony is a topic of passionate debate in birdsong research. Do birds perceive and use the natural harmonic series, or is it only humans who can perceive an octave as the same note? The butcherbird and magpies in Australia are clearly singing notes in the harmonic series, but do they know that, or is it merely an accident? If you listen to slowed down birdsong, it's also clear that most birds are not restricted to whole notes; like Indian musicians, they love the gaps between the notes in the well-tempered scale. But the harmonic series is so fundamental to nature that it would be unsurprising if birds, too, appreciate it. More research is required, because so far nobody knows.

Not all birdsong is pleasing to humans. Starlings' boisterous early morning conversation can be quite challenging to human ears, though there is no doubt that starlings are singing a song with its own complexity and structure. So again, do we define something as song only when we like it? Or do we project this metaphor onto nature in order to connect with something we barely perceive that's deep and mysterious and gives beautiful meaning to a world beyond our control, satisfying the primal need of our species to understand and explain reality?

Of course, not all birds sing with structured performances featuring a beginning, middle and end. Crows, ravens, magpies, penguins, ducks, geese and sparrows use sound linguistically, akin to the way that dolphins, beluga whales or orcas use their squeaks and clicks to communicate with one another. Among the most recognizable and meaningful bird calls is the proud morning *cock-a-doodle-doo* of an alpha cockerel, while the soft, soothing coo of a pigeon has its own rhythmic cadence.

Ofer Tchernichovski, Professor of Psychology at Hunter College, City University, New York, has published studies on birdsong for over twenty years. He has determined that although there are species-specific styles of song, individual birds do vary in ways that may be too small for us to notice but are probably glaringly obvious to others of their kind. This makes singing a better song an important way to capture a mate and propagate an individual's genes, which is the most primal urge in all of nature. Singing well has its rewards, and the competition to be the best singer must be a key driver in the development of more complex or beautiful songs over time for at least some birds.

Songbirds can learn songs by listening to their peers, but experiments have shown that even a bird reared in complete isolation will learn to sing its species' song to a standard that is acceptable to others of its kind if it is released back into the wild. The song is in the bird before it hatches: this is nature, not nurture. In other experiments, birds have been taught the wrong song from birth and then released into their community; quite soon, such a bird works out what the right song is and starts to sing it instead.

This ability to listen to a sound and then reproduce it is known as "vocal learning," as distinct from "auditory learning," where, for example, a dog will learn the command to sit—not a capability apparently shared by most cats, although they may simply be choosing not to learn such things. Laboratory analysis has revealed that the vocal-learning ability in humans and birds involves (though is not exclusively based on) a gene called forkhead box protein P2, or FOXP2 for short. This gene exists in many vertebrates in slightly different forms: it is important in bat echolocation and even mouse vocalizations, but only in humans and birds is it known to be crucial for speech development by vocal imitation. Mutations in this gene in humans are now known to cause ADHD, as well as developmental verbal dyspraxia (DVD).

Vocal learning is rare in nature—primates other than humans do not possess it, which is why, despite their intelligence, we can't teach chimpanzees to speak and must rely on sign language alone—but it's not confined to humans and birds. We will shortly meet singing whales and dolphins with individual names, while bats, seals, sea lions and elephants can also learn sounds and then vocalize them. It's beginning to emerge that this capability is not just present or absent in the animal kingdom: there is a continuum, from no ability to human. More research is going to be needed before we fully understand how vocal learning works in the brain and why some intelligent animals with sophisticated language capability, like chimpanzees, simply can't do it.

In her book *The Sounds of Life*, the late Professor Karen Bakker explored how artificial intelligence is being deployed in an attempt to decode animal communication. One example is the Cetacean Translation Initiative (CETI), which has deployed arrays of buoys and drones in Dominica to record the clicks of sperm whales and is using AI to identify the equivalent of syllables, words and syntax in the expanding database of sounds. The ultimate goal is to learn how to speak to, as well as understand, these extraordinary animals, which may well be at least as intelligent as humans. CETI is a pathfinder in this field, and this AI-based approach is undoubtedly going to inspire fascinating research into the communication of many other species, from birds and bees to primates and mice. I suspect that when AI draws back this veil, our arrogant presumption that humanity has exclusive rights to consciousness, intelligence, society and language will be demolished, and our species will have an epiphany that will forever change the way we treat the animals that share this planet with us.

Not all bird sounds are songs or social communication. The oilbird of South America forages during the night and sleeps in caves during the day, behaving rather like a bat—and, like a bat, it has learned to use crude echolocation to help it navigate in the dark. Although it produces some sounds in the ultrasonic range, most of its echolocation sound is less than 20 kHz, which is within human hearing and far less directional that bat signals. Cave swiftlets in Java, Sumatra and Bali, whose nests, made entirely from their own saliva, are still harvested to make bird's nest soup, also use this kind of echolocation to find their way in the caves in which they sleep at night.

At the other end of the scale of bird sound is the quietest bird of all: the owl. An owl can, of course, screech when it wants to, but in hunting the aim is silent running—and not just in the human audible range. We now know that the flight sounds of many birds contain ultrasound, so owls that prey on small mammals have evolved special measures to ensure that the ultrasonic hearing of their prey does not alert them to the approach of this otherwise completely silent hunter. They do this with three modifications of their feathers: a fringe on the leading edge, another on the trailing edge, and a downy upper surface. Originally thought to be purely about eliminating audible flight noise, we now know that these adaptations also muffle very high-frequency sound. The fish-eating owls of Asia don't need them because most fish have no high-frequency hearing, so they *do* create ultrasound during flight.

Singing birds like blackbirds, thrushes, wrens and nightingales have complex songs that move incredibly fast. In order to appreciate the wonder of birdsong fully, you really have to listen to it slowed down. Even corvid calls make more sense to human ears at half-speed, and if you listen to a nightingale or thrush at quarter- or even eighth-speed, you start to comprehend and appreciate the sublime musicianship and surreal beauty of what these tiny creatures are doing [13]. Other birds must be able to hear this in real time. It seems that birds hear much faster than we do.

It has also been established that the act of singing causes dopamine to be released in a bird's brain. One could say that birds are addicted to singing, though that seems a reductive and rather empty explanation for why they create such long and beautiful songs.

Birds are definitely listening to each other when they sing, though it doesn't always affect their own singing behavior. David Rothenberg distinguishes three behaviors in nightingales. In the first, a bird politely pauses until its neighbor has finished, and then takes its turn. In the second, possibly where territory is in dispute, a bird will start singing when a neighbor does, trying to outdo the other singer. In the third, a bird will completely ignore others and sing exactly what it wants, when it wants. Most human musicians know someone like that.

A few birds listen attentively enough to collaborate in compound songs where the singers alternate. The Australian eastern whipbird has a call-and-response sound that's the dominant sound in the rainforest: the male sings the iconic whip sound, then the female responds with two quick pulses [14]. This is a sound that will instantly transport many expatriate Australians thousands of miles back to their homeland.

Andean plain-tailed wrens easily trump that simple call and response with their tightly meshed duet that, if you didn't know, sounds like a single individual—and it's fast. Each bird leaves brief gaps for the other to fill, and their timing is impeccable and flawless [15]. Brain scans show that each bird learns the entire duet rather than memorizing only its own part. The female leads with a solid melodic structure while the male contributes with more variable intersections, possibly so that he can show off his creative prowess and win the hand (foot?) of his singing partner.

Donald Kroodsma is professor emeritus at the University of Massachusetts, Amherst, and one of the world's leading experts on the science of birdsong. In analyzing and recording thousands of birds, he found that, though a few birds simply repeat a single sequence, many have an extensive repertoire of songs to choose from. A male Bewick's wren, for example, may have twenty songs in his repertoire. He will repeat each one up to fifty times, then move on to another. Twenty songs would be enough for most bands in concert, but the male brown thrasher goes way beyond such a modest set-list with a well-documented repertoire of over a thousand songs; some researchers say it may be as many as three thousand. That is an incredible amount of data for a tiny brain to hold.

Not only a memory champion, the Bewick's wren is also a mimic, impersonating wood thrushes and northern cardinals, among others. Mimicry is not uncommon in the avian kingdom: there are dozens of birds that can mimic sounds ranging from other birds and animals to speech and the sounds of human machinery.

The reasons for mimicry are varied but little known. Some mimics may be defensive or aggressive, pretending to be something bigger or tougher than they really are—for example, mimids such as mockingbirds, catbirds and thrashers may make the sound of a hawk to scare other birds away from a source of food. Others may just start involuntarily using their vocal-learning capability to mimic sounds they hear every day, something that often occurs in urban soundscapes with car alarms, for example. Social birds like parrots use sound for communication, so learning a language is crucial in their ability to function within their community [16]. Parrots even call each other by name, according to a Cornell study done in 2011, which makes them only the third living thing that we know of to do this, alongside humans and dolphins. Another likely reason for being a brilliant mimic is that primal driver, gene selection again: skill as a mimic is a fair proxy for intelligence and so is likely to have become a selection factor in the mating game.

This last factor is probably the driving force behind the evolution of the world's undisputed master of mimicry, the superb lyrebird—that's its full name, not my accolade. For fifteen million years lyrebirds have been honing their astounding ability to imitate any sound they come across, and for the male of the species it is without doubt a performance that's designed to impress—though the fact that females, too, are adept mimics tells us that sexual selection is not the only reason for their behavior.

After clearing a suitable bare patch as a stage for himself in the Australian rainforest, the male superb lyrebird will display his impressive tail plumage to best advantage and start to reel off his finest impersonations. Up to eighty percent of his vocalizations will be impressions, and these are among the most jaw-dropping sounds in all of nature. There is a famous scene in David Attenborough's *Life of Birds* [17] where a male produces a series of uncannily perfect imitations: a kookaburra impersonation so accurate that a nearby genuine kookaburra starts to respond; a camera shutter with and without motor drive; a car alarm; and even a stunning rendition of a chainsaw that it must have heard during nearby logging. It's interesting to speculate how these birds communicate outside of courtship rituals. It would be exhausting to live with a comic who can't stop telling gags, or an impressionist who is always someone else. I wonder who the real lyrebird is, and how much of the time he or she surfaces from the sea of mimicry.

Aside from the desire to propagate and the sheer joy of aesthetic creation, another key driver in the development of complex birdsong is environment. All animal sound is a product of its environment to some extent, and birds certainly have not evolved in isolation. They sing in, and as part of, a sonic context. There are very few occasions when biophony is cacophony. Any sound-maker will adapt to ensure that it can be heard and identified, unless (as with frogs or crickets singing in unison) the opposite is required for self-preservation.

There is research to support the idea that environment impacts animal sound. Ducks in the UK have developed regional accents because of the background noise they experience. Comparing ducks at Spitalfields City Farm in east London to those at Trerieve Farm in rural Cornwall, a 2004 Middlesex University study discovered that the London ducks were louder and more raucous than their country cousins. "The cockney quack is like a shout and a laugh, whereas the Cornish ducks sound more like they are giggling," commented Dr. Victoria De Rijke, who quickly rejoiced in the sobriquet Dr. Quack when the survey made the news. Similarly in Berlin, where David Rothenberg has done hours of recording with the many nightingales that live there, ornithology researcher Henrik Brumm found that urban noise forces birds to sing up to

14 dB louder, peaking at an impressive 95 dB, which is comparable to a chainsaw in volume.

In San Francisco, esteemed ornithologist Luis Baptista (a firm believer in the influence of birdsong on human composers) spent years recording white-crowned sparrows and established that birds in different neighborhoods had their own dialects. After his sudden death in 2000, others revisited the task and found that increased traffic noise was now driving the birds to coalesce in a single dialect that was higher in pitch, and therefore more audible against background traffic din. Incidentally, Baptista's work also showed that the sparrows in each area developed their songs to occupy the sonic spaces with the least acoustic energy, supporting the hypothesis that all animals find their own niche in which to make themselves heard.

This is getting harder due to human noise. Waiting at a walk/don't walk sign in Manhattan, I have often looked up to see a single bird on top of the pole, singing at the top of its voice, and yet inaudible from just a couple of meters away in over 80 dB of traffic noise. I always feel despondent at this little drama, where the tiny bird is trying its best but losing the hopeless battle against an implacable and indifferent opponent. It's a poignant microcosm of a global conflict between nature and technology.

This brings us to the best reason for paying attention to birdsong. As David Rothenberg says, "I think taking birdsong more seriously can get people to appreciate nature more, and hopefully work to save it, to live with it, to be closer in harmony with it and not destroy it."

WHAT YOU CAN DO

Birdsong is perhaps the single element of nature that resonates most with human beings, and now we're discovering that it's actually good for us. The University of Surrey's Dr. Eleanor Ratcliffe has researched the effects of birdsong, and her papers show that most people take pleasure in birdsong, especially if it's quiet, high-pitched and harmonic; further, listening to birdsong reduces feelings of stress and, as part of a biophilic soundscape involving other elements such as water, can reduce the physiological signs of stress such as heart rate and skin conductance.

Some of this works through association. We have learned over hundreds of thousands of years that when the birds are singing, we are usually safe. Birds stop singing when danger threatens, as noted, often when a tsunami is on the way—more evidence of their ability to hear infrasound, which would precede such a destructive wave long before it became visible. We shouldn't underestimate birds' function as nature's alarm clock: when the birds start singing, it's time to wake up and become active.

Ratcliffe's research has also revealed that association works hand in hand with familiarity: her subjects rated birdsong that they might hear locally as the most relaxing sound, while unfamiliar bird sounds from other parts of the world were perceived as strange or even threatening and stressful. It was very clear from her work that birdsong exists powerfully and positively in many people's memories: the research found that listening to a bird in the present often unlocks fond memories of childhood and leaves people feeling happier as a result.

For all these reasons, carefully chosen birdsong is an excellent background or masking sound for working: it encourages you to be awake and alert, while simultaneously reducing stress and potentially improving your mood—and it's good for your health.

Terra sonora

Now we come to the land-based animals. It may surprise you to learn that many familiar animals, including dogs and cats, but also horses, cows and sheep, hear far into the high ranges of ultrasound. The reason they can detect such high frequencies is not to hear others of their species (they don't generate sounds that high). Rather, it's so that they can better estimate where sounds are coming from—a skill that's crucial for survival, whether those sounds are made by predator or prey.

This is the same process we have already encountered—localization by use of binaural spectral-difference cues, the tiny differences in timing or volume of a sound arriving at each ear. Remember that low frequencies with their long wavelengths penetrate matter easily, while high frequencies are easily blocked when they hit something solid. If you are an animal with a small head, you need to focus on the frequencies that are effectively blocked by your skull in order to sense that difference in timing. Low frequencies pass straight through small skulls, but the higher a frequency, the more even a tiny skull will block it and create that crucial timing difference as the sound takes time to go around the head and arrive at the far ear.

This is why there is a clear mathematical relationship between head size and high hearing thresholds, as beautifully shown in the graph below from a paper by Henry E. Heffner and Rickye S. Heffner of the Department of Psychology at the University of Toledo that appeared in the *Handbook of the Senses: Audition*.

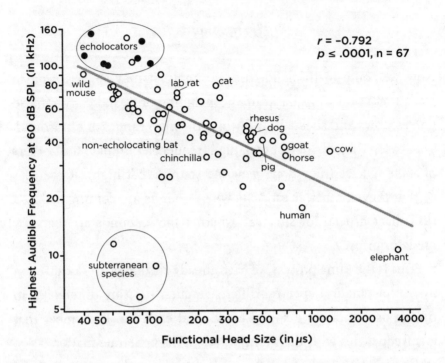

As head size increases (the horizontal axis), so high-frequency hearing (the vertical axis) becomes less necessary to localize sound sources, so the upper limit of the hearing range diminishes.

It's interesting to see that some cave-dwelling species (specifically naked mole-rats, blind mole-rats and pocket gophers) have no high-frequency hearing at all, making them unable to localize sound. Due to the way sound propagates in the burrows they inhabit, almost all the sound they encounter is low frequency (500 to 1,000 Hz) and it can only come from in front of them, so their hearing has adapted and is tuned accordingly. By contrast, echolocating bats (the black dots) and cetaceans (the two white circles in the echolocators ring) can hear rather higher than the regression line would predict, because they are actively using these frequencies for hunting and navigating.

The high end

Bats deserve deeper investigation because they are probably the best listeners on our planet. Masters of the high-frequency end of sound, bats are thought to be the second most common type of non-human mammal on Earth after rodents, and the only ones that can fly. There are over a thousand species of bat, and they live on every land mass except for Antarctica. They often congregate in enormous numbers: for example, over twenty million bats converge to breed every summer in a single cave near San Antonio in Texas.

Most bats are nocturnal, so their daily challenge is to catch and eat their own bodyweight in insects—in pitch darkness. Their secret weapon is echolocation, or biosonar. What they achieve with it is frankly astounding. With brains about the size of a peanut, they can interpret the returning echoes of their high-frequency calls precisely and quickly enough to distinguish a fast-moving moth from another bat or a tree, and to vector in at high speed and catch it—but not with their mouths.

Seth Horowitz, whom I had the pleasure and privilege of knowing before his untimely passing, was a neuroscientist who spent years researching bats. Seth described the extraordinary hunting maneuver of a bat to me: "They're sending out a chirp, they're waiting for the echo to come back and, as they're approaching a potential target, they chirp faster and faster and they're getting the echoes back quicker and quicker… when it reaches the point of no return, where the bug is a couple of centimeters right in front, they stop chirping and they do a kind of a back flip. It's not that they catch it in their mouth. Most echolocating bats will take their tail, raise it up and flip the bug into their mouth. They do this kind of half-somersault, swallow, recover and then continue chirping." This all happens in a fraction of a second, because a bat can't afford to fly blind for longer than that. Its mouth is in effect its eyes, and it must get back to chirping as fast as possible.

Flying at up to forty kilometers an hour, bats can pull 9G as they turn, in pitch-dark environments that are full of complex obstacles. It is no exaggeration to say that they must be the world's most attentive listeners because they see in sound, and one moment's inattention could be fatal. But how can these tiny brains possibly create and update a real-time 3D map of their surroundings fast enough to let them fly and hunt like that? Seth discovered that they have a unique adaptation in their brains: a filter at the lowest part of the auditory system that allows only the first signal from any frequency to be processed, reducing noise and neural load and making their audio processing instantly perfectly accurate.

Bats are extraordinary in several other ways. They have a very fast metabolism: while hunting, their heart rate can exceed a thousand beats a minute, but some species can slow their heart down to just four beats a minute when hibernating. They can live for more than thirty years, and it seems that they have solved a problem that plagues all other mammals, including humans: presbycusis, or age-related hearing loss. Presbycusis doesn't seem to be about wear and tear: even mice suffer degradation of their high-end hearing, and they live for only two years at most. Whatever the cause, bats do not suffer from it: they never lose the high end of their hearing, which would, of course, be fatal to them as they would be unable to hunt. Scientists are ardently studying bat hearing to discover their secret, which just might help us to restore perfect hearing to up to half of the world's population.

Outside of echolocation, very few animals produce ultrasonic sound. Some rodents do, which is another good reason for cats to combine their great sensitivity and movable outer ears with the ability to hear acutely far into the ultrasonic range: it means they can hear rodent communication. Since the 1950s it has been known that some myomorphs (mice, rats, hamsters, voles and lemmings) can vocalize in ultrasonic squeaks. A 1956 study found that baby mice removed from the nest produce calls at 70 to 80 kHz, around two octaves above human hearing—which, probably not coincidentally, is the exact upper limit of cat hearing. Adult mice go ultrasonic mainly during intense experiences like mating or conflict. But, as we discovered with insects, very high frequencies, though highly directional, are easily blocked, which limits the range of ultrasonic communication in habitats that include trees, grass, walls or other obstacles. To communicate over distance, mice have to lower their range and use frequencies that even humans can hear.

Frogs, too, venture into ultrasound. Torrent frogs in China and Java and Borneo's charmingly named hole-in-the-head frog are the only two non-mammalian animals known to vocalize with strong harmonics in the ultrasonic, and to be capable of hearing these sounds. Only the male frogs can call in ultrasound. They have larynxes just half the size of the females, and specially adapted ears with thin, recessed eardrums that enable them to hear up to 35 kHz. They may have evolved this ability in order to make themselves heard over a background of fast-flowing water or waterfalls, which is their natural habitat. As the females can't create and probably can't hear ultrasound, the males' singing ability must be about territory: each male has a patch on the bank of a noisy stream around 1.5 meters from his neighbors on either side, and the songs help the frogs to maintain the correct distance. Their songs are not unlike birdsong or whalesong, but they do not repeat: each warble is different. Those frogs are very unusual with their high pitch, but frog vocalization is almost universal. They've had over two hundred million years to develop their sounds, and there is a stunning diversity across the species, matching the variety of their chosen habitats, which range from fresh water to a life permanently underground or high in the forest canopy, where some can even glide from tree to tree. Frogs have larynxes, and they make the most of them. The rather parodic *"ribbit"* sound is not at all universal. Frogs creak, whistle, squeak, quack, chirp and bellow, and they run the gamut of frequencies from ultrasonic to the low groans of bullfrogs [18].

Frogs are excellent examples of the value of sound for so many living things. Like a cicada or a snapping shrimp, a tiny frog can make itself effectively thousands of times larger by creating an outrageously loud sound that can be heard throughout a sphere as much as a hundred meters in diameter. The common coquí frog—much loved in Puerto Rico but now considered an invasive pest in Hawaii, California and elsewhere—is just three centimeters long but the "*ko-kee*" two-tone call that gave the frog its name can be as loud as 90 dB at one meter. The lower first tone is aimed at rival males, and the higher second tone at attracting females; the whole call sounds more like a bird than a frog and is probably the sound most associated with Puerto Rico [19].

Frogs can even win awards with their sound. A few years ago, I teamed up with the website beautifulnow.is to run an online competition for the world's most beautiful sound. Nature sound recordist Bernie Krause and I drew up a shortlist of nine sounds from the many entrants, ranging from giggling babies to purring cats, waves on a shore (many versions!) and chimpanzee vocalizations. The winner, chosen by a vote by the Beautiful Now subscribers, was a superb recording made in Borneo by Australian nature sound recordist Marc Anderson. It's called "Dusk by the Frog Pond" and stars the *"wrooerks"* of numerous mahogany frogs against a backdrop of empress cicadas and a few late-lingering birds [20].

Like crickets, male tree frogs form choruses that make it hard for predators to localize any individual, using confusion and sensory overwhelm as a form of defense. And like crickets, these choruses often fall into sync to complete the auditory illusion [21]. Not calling at all would be safer, of course, but that's not an option because this is how they attract females.

Nature's bass players

Infrasound is sound that's too low for us humans to hear, which means anything below 20 Hz. Whales, elephants, hippopotamuses, rhinoceroses, giraffes, okapis, alligators and even peacocks make sound in this low range.

The sound most associated with elephants is trumpeting, but much of the time their output is far lower and less strident, and it can travel great distances. Elephants rumble at frequencies as low as 12 Hz, and they can do it very loudly—over 110 dB at one meter—which means that this sound can travel several kilometers, because airborne low frequencies suffer very little excess attenuation from the environment: they just go straight through most obstacles [22]. Elephants have huge middle and inner ear parts and their cochleas, like those of reptiles, specialize in sensing vibrations at low frequencies. As a result, they can hear sounds that are octaves below human perception, although their high-frequency hearing is limited to 12 kHz (I know how that feels, because that's the top end of my hearing now after years hitting cymbals).

Elephants also listen with their feet. They produce infrasonic seismic waves that travel through solid ground and can be sensed many kilometers away by other elephants using bone conduction and Pacinian corpuscles in their feet, which they plant strongly, causing them to splay out, when they are listening to the ground. The potential range of this communication is not known, but infrasound can travel further and faster in rigid solids than in air. The experience of Dame Evelyn Glennie learning how to perceive sound through tactile vibration shows how close hearing is to touch, so this skill may be latent in all human beings too; perhaps we've just become so reliant on our ears that we've become numb to a much wider range of sounds that we might be able to perceive with our hands, feet and bones, even if our ears are not configured for the task.

The almost-extinct Sumatran rhinoceros is another highly vocal user of infrasound. Much smaller than the African rhino, this is a completely different genus that split from other rhinos over twenty million years ago. Hunting and habitat destruction have brought them to the brink of annihilation: there may be as few as eighty in the wild, and in Malaysia there are just three left. Desperate efforts are being made by Sumatran Rhino Rescue to save the species. Friendly vegetarians, they make three distinct sets of sounds. One is a high-pitched vocal squeak or whine that tends to indicate a request or greeting and can be sequenced into a long vocalization almost in the style of a humpback whale [23]. Then there are breathing patterns, mainly panting, indicating anything from anxiety to reassurance.

Finally, they have their superpower: a whistle followed by a burst of air creating a powerful infrasonic sound that can resonate thick iron bars and will carry almost twenty kilometers in the animal's densely forested natural habitat.

Perhaps the most surprising user of infrasound is a refugee from our discussion of birds. In 2015 Angie Freeman from the University of Manitoba, Winnipeg, published a paper revealing that the dazzling visual impact of a peacock flaunting his iridescent plumage is not all that's going on: the display also produces infrasonic signals, which both male and female peafowl can hear. Perhaps the cross-modal impact of sight plus sound helps a peahen to choose the best mate.

Also at the low end of animal sound we find reptiles. Snakes lack both outer ears and eardrums, so their jawbones act as coupling elements to pick up ground-borne infrasonic vibrations and deliver the acoustic information to a cochlea-like mechanosensory system. However, their vocalizations are generally limited to warning hisses.

Lizards have rudimentary ears and hear mostly in a very restricted midrange, while also being sensitive to infrasonic ground-borne vibrations; most have no vocal cords and make no sound, though some do hiss. The exception is the gecko, which vocalizes extensively with clicks, chirps and squeaks [24] or even, in the case of the leopard gecko, an extended high-pitched scream when disturbed.

Larger reptiles have a much stronger relationship with sound. Crocodiles can hear all the way up to 8 kHz, and are most sensitive at 100 to 3,000 Hz. Bull crocs hiss, roar and bellow to threaten other males and to attract females [25], and babies still in their eggs vocalize and respond to the sounds of their similarly encased siblings, which may be how they achieve synchronized hatching. After birth, the juveniles make short barking sounds [26] and it's clear that sound plays an important role in crocodilian parenting and family life. Different species of crocodile have their own dialects, but researchers have established that these share the same basic acoustic code.

Many reptiles conserve their energy with a sit-and-wait hunting strategy, like crocodiles lurking just below water level, and almost all hunt alone, so they have not developed the need for the social communication or bonding that is so evident in the teamwork of lions, wolves, orcas or ants.

Pet sounds

Cats have been recorded making more than twenty different vocalizations. With humans, it seems that domesticated cats favor meowing and purring, but they can also chirr, chirp, growl, hiss and more depending on the circumstances; most town dwellers will be only too familiar with the drawn-out yowling and calling that can accompany cat interactions in nighttime gardens [27].

The mechanism of purring is still something of a mystery: cats large and small (and their close relatives the civets and genets) are the only animals that truly purr. Although plenty of other animals, from bears and gorillas to rabbits and kangaroos, can make a similar sound, this is not a genuine purr. Even among cats, only some have the ability: any cat that can roar cannot purr, and vice versa.

Cats can certainly communicate their needs to humans quite well, but it seems to be almost entirely one-way traffic in most homes, and there are few cat obedience classes, possibly because most of them don't subscribe to the concept. However, cats are in fact trainable and there have been performing domestic cat shows since the nineteenth century, not to mention the much riskier exploits of circus lion and tiger trainers.

On the subject of the non-purring cats, no discussion of biophony would be complete without acknowledging possibly the world's most iconic animal sound: a lion's roar. Intriguingly, most people's idea of this sound is a fabrication. MGM have been using a lion as their logo since the silent movie days of 1916, so when sound became possible in 1928, they added a recording of the actual lion pictured. This was updated in the 1930s and then used without change until 1982, when movie sound designer Mark Mangini was commissioned to create a new sound in stereo. He used multiple tiger sounds mixed together to create a sound effect that is undeniably majestic and ferocious—but definitely not a lion's roar. You can find examples of the evolution of this iconic sound on the internet.

The real thing is used in the wild by both male and female lions to warn off intruders into their territory, and with its low-frequency elements can carry long distances: it's estimated that lions as distant as ten kilometers away may be able to hear a good roar.

Roaring takes some vocal modifications, which is why humans can't do it: it requires an enlarged larynx, ideally one that can be pulled deep into the throat using special muscles; extra spaces in the throat or head to add depth and resonance; and, in the case of the big cats, square vocal cords (those of purring cats are triangular).

Lions are not the only animals to carry off an impressive roar. Along with the other big cats, roarers include bears, elephants, great apes, elephant seals and red deer. I grew up in a house backing onto

London's Bushy Park, and the raucous, guttural bellows of the rutting stags always coincided with the first chill nights as summer departed. It's a noise I deeply associate with heavy dew, fallen leaves and the russet colors of autumn sunsets. Plenty of large male deer around the world emulate the red deer in roaring when it comes time to challenge other males and attract a harem of females [28].

Most dogs, in contrast to cats, do readily respond to human commands, vocal and otherwise, possibly because their brains contain twice as many neurons as cat brains. The apotheosis of human–dog communication is found in sheepdogs. Mainly border collies, these dogs combine their own keen intelligence with complex commands from the shepherd, given in voice or coded whistles, to carry out incredibly complex maneuvers. Watching a skilled shepherd with one or two dogs working a flock of sheep is by turns jaw-dropping and inspiring.

Whistled commands for sheepdogs are potent communicators, but even richer are the human whistled languages, for example *kuş dili*, or Turkish bird language, that evolved all over the world to allow distant communication outdoors but are now all on the verge of extinction. Mobile phones make these human whistled communications obsolete, so shepherds will almost certainly be the only ones communicating in whistles in a few years.

Dogs, like cats, have a range of vocalizations that vary from breed to breed. We had a lovely Samoyed called Ice, who personified the primary function of his breed—to be a watchdog—and expressed it in the most familiar dog vocalization: if in doubt, he barked. If anyone came within his very sensitive earshot of our house, his urgent barking made sure that his extended pack were all aware of the intruder. His barking redoubled when he was excited, especially when a walk was in the offing (we had only to go near the boot cupboard and he was off), and he was loud—well over 100 dB if you were unfortunate enough to be within a meter. I took to wearing headphones in the car we drove to our favorite beach walk in order to damp down the ear-splitting noise. He died in 2023, and though our lives are much quieter now, we miss his exuberance and endless enthusiasm.

The rest of our pets are a very different type of sled dog: Siberian huskies, who are much closer to wolves and exemplify the strong, silent type. Huskies almost never bark, though they may growl or howl from time to time; in general, they take what they want, wherever it is and whoever it belongs to, without seeing the need for any vocalizations.

Wild cousins of domestic dogs can make eerie sounds, as I discovered many years ago. I was standing with a friend at the edge of the Californian desert as the last flame-orange rays of the setting sun were lighting up the tops of the scrub-flecked rocky hills on either side of me. Suddenly, an unearthly wailing emanated from the boulders on the hillside right next to us; it sounded as if a child was in desperate distress, or an animal being tortured [29]. I could see nothing there. Within seconds, it was joined by another, and another, and another from the other side: we were surrounded. To hear coyotes in the wild was uncanny and a direct connection to an ancient and frightening sonic experience—the sound of predatory pack animals surrounding you. We left the scene at some speed, though we were later assured we had not been in any danger.

Apes and bears

Although primates can't manage vocal learning, many make sound and can learn the meaning of sounds, including human speech. Silverback gorillas can generate an intimidating roar, and many apes and monkeys communicate vocally, from grunts and whistles to shrieks and the "*hoos*" of chimpanzees [30]. Much of this sound is audience-related and occurs in the context of social interaction, be it danger warnings or food-related calls, bonding, mating, disagreements or fighting. Like humans, primates generally have good hearing with a broad range of frequencies, and most can hear higher frequencies than we can, mainly because they have smaller heads, which as we now know require higher-frequency sensitivity if the animal is to be able to locate the direction from which a sound is coming.

Two sonic outliers must be mentioned. The howler monkeys of Central and South America are the vocal champions of the primate kingdom, and some of the loudest animals on Earth. They have unique adaptations in their throats to achieve their extraordinary volume: their hyoid bone is extended, and there are special air sacs that amplify their calls, making them audible almost five kilometers away in dense jungle. Before dawn, they groan with a deep sound not unlike Tibetan monks chanting with undertones, in order to establish each individual's territory for leaf-gathering that day. The noise becomes deafening if two troops meet, when males indulge in a sing-off to establish supremacy [31].

More aesthetically pleasing to human ears, but equally overbearing, is the sound of the lar gibbon, whose natural habitat is the dipterocarp rainforest of the Far East, once including southern China and the entire Malay peninsula but now much smaller. At rest, these brilliant brachiators emit a tuneful short midrange upward swoop, which can build in intensity to become a glorious, deafening upward glissando ending at notes a soprano would be proud of. These musical apes can also perform intense trills, and some of their sinewave-like musical sounds could be lifted from a Keith Emerson Moog synthesizer solo from the 1970s. Breeding couples create their own variations on the family theme, which is sung with great gusto and at extreme volume every morning to establish tribal territory [32].

Like primates, many types of bear are social and have strong family groups that persist for well over a year, so social communication with sound is very common. Bears have good hearing combined with high levels of intelligence, so they are natural users of sound.

Though there is very little research on bear hearing, one 2007 paper measured polar bear hearing and discovered it to be excellent in the mid to high frequencies. Polar bears hunt seals in the Arctic pack ice (for as long as that exists, which is only until 2035 at current depletion rates). It seems that they locate their prey through two meters of ice by careful listening, making out the calls of ringed and bearded seals and tracking them to their breathing holes, where they are most likely to find pups or even grab an adult.

Most bears don't use their voices to bellow or roar unless in combat or under threat [33]; in their families, their audible communication is limited to clicking with teeth or tongue, huffs that are like heavy sighs, grunts or pops.

Deep and meaningful

There is no clear correlation in the living world between size and volume: frogs and cicadas show us this with their stupendous noise-generating abilities, while gentle giants like anteaters and giraffes are quietly spoken. There is, however, a strong relationship between size and the depth of an animal's vocalizations: bigger things make deeper sounds. This means it's wise to take deeper voices more seriously because, in general, the bigger the animal, the more potential threat it poses.

This eons-old genetic lesson is why research shows that people tend to vote for politicians with deep voices, and why Margaret Thatcher employed Laurence Olivier's voice coach, Kate Fleming, to help her lower the pitch of her speaking voice by 46 Hz, according to Max Atkinson's 1984 book *Our Masters' Voices*—several tones, and almost half the average difference between male and female voices. She knew that we associate deeper voices with greater importance and, by inference, authority.

WHAT YOU CAN DO

If you want to speak with greater authority, it can pay to lower your pitch. Here's a simple way to practice. Place a finger or two on your sternum and read some text, focusing on resonating in your chest so that you can feel your fingers buzzing with the vibration cause in the chest bone. This feedback loop is the key: regular practice will help lower your voice.

*

There is no linear relationship between animal size and hearing range: for example, both elephants and goldfish can hear very low sounds. The logarithmic chart below shows the hearing ranges of various animals in both Hertz and number of octaves (remember, an octave is a doubling of frequency). The prize for the largest hearing range goes to the ferret, at over eleven octaves.

Animal	Lowest (Hz)	Highest (kHz)	Range (octaves)
Tuna	50	1.1	4.5
Chicken	125	2	4
Goldfish	20	3	7.2
Bullfrog	100	3	4.9
Catfish	50	4	6.3
Tree frog	50	4	6.3
Canary	250	8	5
Cockatiel	250	8	5
Parakeet	200	8.5	5.4
Elephant	17	10.5	9.3
Owl	200	12	5.9
Human	31	19	9.3
Chinchilla	52	33	9.3
Horse	55	33.5	9.3
Cow	23	35	10.6
Raccoon	100	40	8.6
Sheep	125	42.5	8.4
Dog	64	44	9.4
Ferret	16	44	11.4
Hedgehog	250	45	7.5
Guinea pig	47	49	10
Rabbit	96	49	9
Sea lion	200	50	8
Gerbil	56	60	10.1
Opossum	500	64	7
Albino rat	390	72	7.5
Hooded rat	530	75	7.1
Cat	55	77	10.5
Mouse	900	79	6.4
Little brown bat	10030	115	3.5
Beluga whale	1000	123	6.9
Bottle nose dolphin	150	150	10
Porpoise	75	150	11

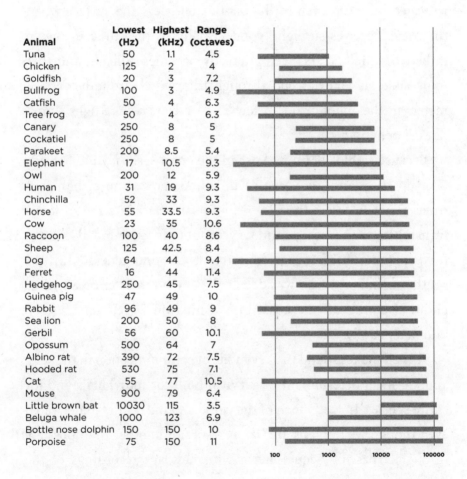

It's time to visit some of nature's most impressive singers. And to do that, we leave land and air behind and journey underwater.

The songs of the sea

Without optical aids and on a clear night, the furthest object you or I can see is V762 Cassiopeiae, a supergiant star that's over 100,000 times more luminous than our Sun and 16,308 light years from Earth. In water, however, even in the photic zone near the surface where the ocean receives sunlight, good visibility is measured in meters rather than light years. Seeing quickly becomes difficult, and then impossible, as you descend through the twilight zone below 200 meters to the aphotic zone below 1,000 meters, where light simply doesn't penetrate.

Because sight is of limited use in the oceans, sound and hearing take up the slack. As we've seen, sound moves more than four times faster in water than in air, and salt water is an even faster transmitter than fresh. Sound travels not only faster but also much further underwater: in one experiment, a sound generated in the southern Indian Ocean was detected some hours later by a hydrophone listening off the coast of Washington State, some 14,000 kilometers away.

This all makes sound the key mode of communication in the seas, and a critical determinant of survival both for individuals and for entire species. In fact, much of the world's sound occurs underwater, even though we humans are barely aware of it, other than possibly through a passing acquaintance with new age recordings of the songs of humpback whales. The oceans are alive with sound, made by almost every organism living there. At the time of writing, the World Register of Marine Species [34] lists over 240,000 species. This is accepted to be just a fraction of what's thought to be as many as ten million species in total.

If you've ever submerged your ears, you will know that hearing is a different challenge in water. As we discovered earlier, your ears transmit airborne vibrations arriving at your outer ear via the eardrum and ossicles (tiny bones) in your middle ear to much higher-density fluid in your inner ear. If you fill your outer ear with water, the whole mechanism is compromised because the pressure differential is virtually eliminated. Despite the much greater speed and range of sound in water, our ears are simply not designed for that environment: we can certainly hear, but it's muffled, and it becomes very hard to detect where the sound is coming from. So how do underwater species hear?

Crustaceans like crabs, lobsters and shrimps don't have ears as we understand them, but it is generally accepted that almost all of them do sense sound, or at least vibration, through various pressure-sensitive structures, including hairs around their bodies. Sensitive hairs extruding at various angles allow an animal to get a sense of the direction and force of any vibration it encounters. This is a sensory strategy that turns up in many domains, including insects, plants and even single cells.

Fish hear rather better than crustaceans. Most fish have evolved a structure called a lateral line that runs along their back and detects changes in pressure, which is a very basic form of hearing. Many have also developed an inner ear, though without either outer or middle ears. The main challenge for fish is that they have almost identical acoustic impedance to the surrounding water, so soundwaves effectively pass straight through them. Fish have evolved an ingenious way of nevertheless detecting soundwaves: inside their inner ear is an otolith (ear stone) which is much denser than the surrounding tissue. You and I hold still while sound vibrates our ear; a fish's ear holds still while sound vibrates the fish, bending nearby hair cells in an organ called a saccule and sending neural signals to the brain for decoding as sound.

Ingenious this may be, but for most fish it's a lo-fi mechanism with quite restricted range and sensitivity (mainly at low to middle frequencies). However, some fish, especially freshwater varieties like goldfish, have a swim bladder full of air that's connected to their ear, allowing them to hear much better. Clupeiformes such as herring and anchovies extend the swim bladder into their skulls and can hear higher frequencies than we can, helping them to better locate sound sources.

Fish that live in shallow, fresh water tend to have the best hearing, while those that live in noisy places like the open sea or fast-running rivers hear less well, probably to avoid being over-stimulated or even overwhelmed by the constant noise around them.

Intriguingly, what many fish are primarily sensing is the motion of particles as they move backward and forward, rather than the changes in sound pressure that oscillate human eardrums. We don't sense or measure particle motion much if at all, which means we are almost entirely ignorant of the effects on vast numbers of marine lifeforms of the sounds we are making underwater, from the cavitation of ship propellers to the explosions involved in prospecting for oil and gas. More work needs to be urgently done in this area.

At the top of the tree of underwater hearing are the aquatic mammals. Amphibious mammals like seals, sea lions, walruses, sea otters and even polar bears need to hear in both air and water, which is no mean task. They have the same ear structure as we do, with outer, middle and inner ear all present, but some have adaptations depending on the amount of time they spend in water. Polar bears and sea otters hear much better in air than in water; eared seals like sea lions have adapted to improve their underwater hearing, and the phocids (true seals) are the most adaptable of all. They have evolved valves that close their ear canals when they dive, as well as huge ossicles (those three bones in the middle ear) and an enlarged eardrum, which allow them to hear up to six octaves underwater, which is comparable to fully aquatic mammals, while still hearing perfectly well in air.

Those fully marine mammals such as cetaceans (whales, dolphins and porpoises) and sirenians (manatees and dugongs) have no functioning outer ear at all, and their ear canals are disconnected from their eardrums. They hear by using vibrating fat and bone conduction in the jaw or the skull, or both. Once sound reaches the inner ear, the process is the same as for you and me, though the specifications of the equipment involved may be very different. Cetaceans use more nerve cells for hearing than terrestrial mammals, and their basilar membranes are tuned to the frequencies they need to pay most attention to. The baleen whales, including the blue and the famously tuneful humpback, have extremely broad thin elastic basilar membranes that allow them to hear ultralow frequencies. Toothed whales have stiffer and thicker basilar membranes than baleen whales, because they need to hear higher-pitched sounds like the clicks they often communicate with. Dolphins continue the upward trend: the orcas that I often see in the waters around Orkney are most sensitive at 12 kHz; common dolphins between 40 and 80 kHz (that's up to two octaves higher than human hearing) and harbor porpoises at 100 kHz. Dolphins' hearing range is enormous, starting at the same 20 Hz as humans but extending up to 150 kHz, around three octaves above any human.

So, what are all these marine animals listening to? Sensing and interpreting sound is crucial for many marine species, whether to interact with others of the same species, to locate food by the noise it makes, or to avoid becoming food for a predator by hearing it coming. Sound can also be a way-finder. Arctic ringed seals make and maintain holes in the ice so that they can access the ocean to hunt and then return to breathe and to feed their pups, which also have an escape route should a polar bear come calling. Adult seals have been known to travel as far as two kilometers under the ice while hunting, then return to their exact breathing hole. Listening carefully for acoustic cues, possibly as subtle as the sonic profile of water moving in their particular breathing hole, may well be a significant part of their navigation method, especially at night.

The oceans are often surprisingly noisy places. There are sounds of movement, which can be very loud: a large school of fish in a feeding frenzy produces hydrodynamic sounds that other fish such as sharks can hear, attracting them to join in the feast. There are sounds of eating: many fish eat the skeletons of crustaceans or corals, and just like a human munching on a biscuit, that can be a noisy business. And then there are intentional sounds made to communicate, whether in courtship, for territorial reasons, to warn of danger, or for social reasons. Much of the intentional sound in the world's oceans, lakes and rivers does not come from vocalizations: many species lack a voice and have evolved creative and sometimes extraordinary ways to create sound without one.

Fish and crustaceans fall into this category. In the absence of a voice, the twenty-five thousand known species of fish have evolved ingenious sound-generating strategies, largely depending on the environment they inhabit. Even when not eating, sunfish and mackerel grind or snap their teeth, while many types of fish make sound by vibrating their swim bladder, or by using it to amplify sounds they make with other parts of their body, including stridulation (the rubbing of the body parts together) in the fashion of crickets or grasshoppers. A famous example is the Californian singing fish, or plainfin midshipman, which migrates from deep water to coastal shallows to spawn. The males build nests and then start humming for hours at a time at about 79 to 105 Hz (that's around E2 to A2 if you're musical) to attract a mate. They do this by rapidly contracting a special set of sonic muscles to vibrate their swim bladders, and their noise is often loud enough to annoy local human populations [35].

Fish sounds are generally fairly low frequency: they are typically well below 800 Hz, with fundamental frequencies of 25 to 250 Hz— lower than the range of a human male bass singer. Larger fish, or those that swim in deeper water, tend to make lower sounds than smaller or shallower-dwelling ones. Many types of fish are social, swimming in schools or living in reef-based communities, and it's very likely that most of their sounds are social communication of some kind. For more solitary fish, sound may attract a mate, delineate a territory or express alarm or form some sort of defense in the face of danger.

Crustaceans can be a raucous bunch. Spiny lobsters, crayfish and many crabs rub body parts to make a squeaking sound [36]. Other crustaceans buzz, hiss or even growl. Even barnacles get in on the act, producing pulses of sound with peak amplitudes of around 70 dB as they eat. Most crustaceans are not social animals, so the bulk of these sounds are probably territorial defense or mating calls.

Dwarfing all these in the noisemaking stakes are over four hundred species of shrimp that make snapping sounds as loud as 190 underwater dB, or around 130 dB in air-based measurement.* That's rock-concert loud! These tiny animals, ranging from a few millimeters to a maximum of five centimeters long, are almost certainly, pound for pound, the world's champion noisemakers, beating even those noisy cicadas [37].

Each little shrimp has one oversized claw. It makes its impressive noise by snapping this claw shut in less than a thousandth of a second, creating and then collapsing a large cavitation bubble. In 2001, scientists discovered that as this bubble collapses, it creates a short, intense flash of light, which means the temperature inside the bubble at that moment must be almost 5,000°C—very close to the temperature of the surface of the Sun.

The point of the exercise for the shrimp is that its snap generates a high-velocity jet of water that can stun prey such as worms or small fish. The jet can also carry pheromones to attract a mate, which can then navigate back along the jet if the chemicals are enticing enough. Collectively, the snapping is also useful to other species, which can hear it from up to ten kilometers away and then use it as a navigational aid to find their way back to their home reef. The shallows around the shore are some of the richest sonic environments on Earth.

Unlike cetaceans and fish, marine mammals do have both vocal

---

* As we discovered earlier, decibels in air and in water are calculated from a different base; you have to subtract about 60 dB from an underwater measurement to arrive at a rough equivalent in air-based decibels.

cords and voices, and they can use them underwater, often augmented by other sounds such as whistles, knocks, clicks, grunts and squeaks. Pinnipeds (seals and sea lions) are highly vocal in air: anyone who has visited Pier 39 in San Francisco will have experienced the cacophony of barking from the large community of up to nine hundred "sealebrities" there. They also vocalize extensively underwater with bell-like sounds, trills, warbles and whistles. The legendary nature sound recordist Chris Watson says that the most beautiful animal music he's ever heard is the singing of bearded seals he recorded with a hydrophone dropped into 800-meter-deep water under the sea ice, close to the North Pole [38].

Whales

Of course, the star vocalists of the marine soundscape are the cetaceans. Intelligent and highly social, whales, dolphins and porpoises have evolved complex sonic communication. As we've discovered, odontocetes (more than sixty-five species of toothed whales including orcas, sperm and beaked whales, dolphins and porpoises) can hear best at high frequencies, as that is where they tend to make sound. With a complex system of air sacs and special soft tissues that vibrate as air moves through their nasal passages, they also have a lipid-filled sac in their forehead called a melon that allows them to focus their calls into a penetrating beam of high-frequency sound. They can move into hypersonic ranges: the Guiana dolphin can produce whistles at 40 kHz, an octave higher than humans can hear. Different species of toothed whales use their own distinctive whistles. As well as whistles, odontocetes can produce bursts of sound described as squawks, creaks, bleats, barks, pops or (particularly among sperm whales) clicks [39].

The rather enchantingly named mysticetes (eleven species of baleen whales such as the humpback and blue) use their huge larynxes to make their spine-tingling low-frequency calls. Inside a baleen whale's larynx is a thick ridge of tissue called a U-fold, its own version of vocal cords, which is next to a large inflatable pouch called the laryngeal sac. Baleen whales hold vast amounts of air in their lungs when they dive: a blue whale can take in 5,000 liters of air in a single gulp, allowing it to stay under for fifteen minutes (the longest validated dive) and possibly up to half an hour (the theoretical maximum), with its heart rate slowing to just two beats per minute. Underwater, the whale can push air back and forth between its lungs and laryngeal sac by contracting muscles in its throat and chest, thus vibrating the U-fold to produce sound. By squeezing its laryngeal sac to change its shape, the whale can alter the pitch and volume of the sound it makes.

This sounds quite technical, but the results are awe-inspiring [40]. It is likely that blue and fin whales dive to a layer called the deep sound or SOFAR channel, where low-frequency soundwaves can travel for thousands of kilometers before dissipating, enabling the whales to communicate over vast distances.

One profound discovery made by scientists studying the calls of various cetaceans is that in several of these species, each animal has a unique identifying sound. A bottlenose dolphin has a signature whistle developed in its first year as it listens to the older members of its group—just one aspect of the capacity for vocal learning that these intelligent mammals share with us [41]. Once established, the individual doesn't change its sound. An exhaustive study of the sounds of bottlenose dolphins by University of St. Andrews scientist Dr. Stephanie King established beyond doubt that these animals reply to their unique sound when they hear it, and that they mimic the unique sonic signature of other individuals to call to them, for example when a dolphin calf calls to its mother.

It seems clear that these animals are not homogeneous units in a crowd, like herring or anchovies in a huge school, or worker bees in a hive: like us, they are individuals, known by name in their social groups and possessing personalities. Using their high-frequency calls, bottlenose dolphins can communicate with one another from as far away as twenty kilometers, which is handy when searching for food because it allows a group to spread out and cover much more territory.

Other species are less well researched, but we know that several other types of dolphin, as well as narwhals and belugas, also have unique sonic signatures. It seems reasonable to suggest that all the social cetaceans work in this way.

Most toothed whales live in groups. These vary in size from a handful to many dozens, and studies done on orcas, sperm whales and spinner dolphins indicate that these groups in turn have unique identifying sounds, so that individuals who are far away foraging or exploring can find their way home through call and response, or identify members of their own group when they meet and mingle with another.

But group communication is not just about identification: it's also how many toothed whales communicate in hunting, feeding, mating and other social behavior. Orcas are highly intelligent, skilled pack hunters, whether coordinating a fast dive to produce a large energetic wave that will sweep a seal off its safe perch on an ice floe, or corralling bait fish into a dead end like sheepdogs herding sheep; in all this, they use sound to coordinate their movements precisely with one another [42]. Dolphins use sound or blow bubbles to startle or flush out fish, producing loud pops or even rendering salmon unconscious with bursts of low-frequency sound.

Sperm whales also communicate in groups. Each whale produces a range of clicks called a coda, with information content—be it a name, or some environmental data like the availability of food, or threats—coming from the number, rhythm and timing of the clicks [43]. Nobody has yet been able to decode sperm whale codas, though it does seem that groups share some common sonic DNA, and that some sequences may even identify an individual or a group as part of a much larger clan that may include hundreds of individuals spread over thousands of kilometers.

Sperm whale clicks can serve another crucial function. Along with other toothed whales, some birds and rodents—and, of course, all bats—they echolocate, sending out directed sound and using the reflections they hear back to navigate or identify food [44]. Sperm whales need this because they dive as deep as a thousand meters, where there is no light at all. On these ninety-minute dives they hunt fish and squid, including the giant squid that was semi-mythical until filmed only a few years ago (see Edith Widder's 2013 TED talk for a dramatic video revelation [45]). Thanks to research by Aran Mooney from the Woods Hole Oceanographic Institution in Massachusetts, we now know that at least some squid can hear, but only at low frequencies (50 Hz to 500 Hz) so it's unlikely they would be able to detect and run away from a sperm whale's higher-frequency clicking. It's awe-inspiring to imagine the battles in pitch blackness between a sixty-ton sperm whale with its twenty-six large teeth (all on the lower jaw) and a fifteen-meter-long squid with its sharp beak and serrated tongue.

Beaked whales are the deepest divers of all, so they also rely on echolocation, using clicks that blur into a buzz as they near their prey in pitch blackness [46]. These are remarkable animals. In 2014 off the California coast, a tagged Cuvier's beaked whale was recorded diving to a depth of 2,992 meters; it was underwater for 137 minutes. At these depths, a whale's lungs collapse under the extreme pressure. To survive on one breath for hours under such pressure, a beaked whale will store prolific amounts of oxygen in its muscles and blood, dramatically slow its heart, and completely shut down its kidney and liver. Just like humans, these whales must take care on the way up to avoid dissolved gas forming bubbles in their bodies ("the bends"), which can be fatal. This may well be the cause of some of the mass beachings of dead beaked whales over the last few decades, possibly after military sonar scared the animals into rising too fast.

Dolphins typically eat much smaller things than deep-diving whales, so their echolocation is accordingly highly tuned: using focused ultrasound, they can identify a target the size of a golf ball at a distance of one hundred meters [47].

We've left the best to last: the song of the humpback whale [48]. First recorded (though not recognized) off Oahu, Hawaii, in 1952, sounds described as "squeals, creaks, cries, barks, groans, and whoops" were identified in 1967 by Ocean Alliance founder Dr. Roger Payne as originating from humpback whales. Payne collated his recordings and released an album, which surprised everyone by becoming a multiplatinum-selling international hit; some of its sounds are on the *Voyager I* and *II* probes that are currently heading out of the solar system into interstellar space. The album played a significant role in stimulating the Save the Whale movement, which was instrumental in achieving the whaling moratorium signed in 1982.

In 1983 it emerged that the humpback songs were coming from males moving toward or having arrived at their tropical breeding grounds at the end of their annual migration of up to eight thousand kilometers from the Arctic or Antarctic waters where there exist the huge krill populations that the whales feed on.

Further research has established that whales in the same ocean basin all sing similar songs, and that the songs evolve from month to month and year to year. This kind of song evolution is rare among animals and the reason for it is not known. Perhaps it stems from creative energy, just like human musical evolution. Humpbacks typically sing between three and nine themes in rotation, where each theme comprises a unique phrase that's repeated until the theme changes. The phrases themselves are composed of one or more repeating "units," where a unit is an identifiable continuous sound. A performance can last a very long time: one whale was recorded in 1978 singing for twenty-two hours nonstop. Sometimes the themes are repeated in the same order, but at other times there are variations. Despite the best efforts of scientists to analyze and impose structure on the singing, it seems that the whales are like jazz musicians, accepting some boundaries but playing relatively freely within them.

Despite decades of study, scientists cannot say why the whales sing, other than to suggest it has something to do with mating—an unsurprising observation that applies to a lot of animal sound, including human music. One thing we do know is that ship noise interrupts humpback singing: Japanese research in 2018 found that fewer whales sang within five hundred meters of a shipping lane, and that the whales stopped singing altogether while ships passed by.

Humpbacks are among the few animals that are capable of vocal learning—hearing a sound and then reproducing it. Between 1995 and 1998 a team from the Department of Veterinary Anatomy and Pathology at the University of Sydney found that a population of around a hundred humpback males off the Great Barrier Reef completely changed their song pattern after two individuals joined with a new song that they brought from the Indian Ocean off the west coast of Australia, thousands of miles away. It took just two years before the entire population had switched to the new song, possibly because it was fresh and interesting to them.

In a talk based on his book *Whale Music: Thousand Mile Songs in a Sea of Sound* [49], David Rothenberg notes that speeded-up humpback whalesong is remarkably like the song of the nightingale, and that slowed-down nightingale song sounds very like whalesong. How uncanny that two animals so different in size, habitat and evolution should sing in similar ways. The meaning of this? Nobody knows—but it is humbling to find these connections between very different members of the vast orchestra of animal sound, which is where we now turn our attention.

The great animal orchestra

In this chapter, we have taken a short tour through what eminent nature sound recordist Bernie Krause dubbed biophony. In his bestselling book *The Great Animal Orchestra*, he eloquently describes its enormous range and variety:

> Animals are hooting, bleating, growling, chirping, warbling, cooing. They are tweeting, clucking, humming, clicking, moaning, howling, screaming, peeping, sighing, whistling, mewing, croaking, gurgling, panting, barking, purring, squawking, buzzing, shrieking, stridulating, cawing, hissing, scratching, belching, cackling, singing melodies, stomping feet, leaping in and through the air, and beating wings—and doing it in a way that each voice can be heard distinctly, so that the animals seem to be able to hear and to distinguish one voice from another.

Bernie started his career in sound sixty years ago as a musician. After replacing Pete Seeger in the Weavers, he moved on to become a pioneer of electronic music in partnership with Paul Beaver, working closely with Robert Moog on his ground-breaking synthesizers; from 1967 they introduced the Moog synthesizer to the Doors, Simon and Garfunkel, the Byrds, the Monkees and the Beatles, and scored or played on countless film soundtracks. When Beaver died suddenly in 1975, Bernie changed tracks and set about recording the planet's natural sounds. He went on to amass his enormous collection of beautiful recordings, many of which would be unobtainable now as species have become extinct or habitats destroyed. The trend leaves him deeply concerned. When I asked Bernie what he felt was the future of biophony, and its academic discipline, bioacoustics, he said, "Silence, if we're not very careful."

In five decades of recording nature, Bernie has made over ten thousand sound recordings, covering more than five thousand habitats and fifteen thousand species. His favorite recording was of the explosive exhalation of an orca as it echoed off the granite cliffs in Johnstone Strait, Vancouver Island—a recording sadly lost when Bernie's beautiful Glen Ellen home was burned to the ground in the deadly Nuns wildfire in 2017. His most loved sounds to listen to are "the early morning song of a hermit thrush, with the reverberation of its singular voice reaching to infinity—or a cat's purr." But when I ask him about his favorite place, his answer is tinged with sadness: "The Yukon Delta, Alaska, because Alaska is about three times the size of France and there are only about 750,000 people there. It has a large coastline. It has fabulous interior habitats. It has subarctic rainforest, boreal forest, tundra, mountain habitats. But Alaska has changed radically. The temperature has gone up by an average of four or five degrees Celsius, whereas in most places in the world it's around half a degree. The tundra is melting, and certain species of birds, never before found north of Fairbanks, located at the mid-point in the state, can be found migrating as far north as the Beaufort Sea, nearly four hundred miles (640 kilometers) away. The indigenous first peoples of Alaska living in that area have no name for the bird in their language as it is so new."

In analyzing the spectrograms of countless habitat recordings, Bernie came to realize that animals in any healthy habitat form an effective orchestra, with each species sonically occupying its own niche in space and time and avoiding the acoustic domains of others. The niche thesis became his book *The Great Animal Orchestra* in 2012, a captivating TED talk [50] and then in 2014 Bernie collaborated with UK composer Richard Blackford to crystallize it in a symphony, commissioned by the BBC, for a real-life orchestra. Made possible as a performance by the advent of sampling keyboards, *The Great Animal Orchestra Symphony* mixes the sounds of traditional orchestral instruments with recordings of animal vocalizations, starting with the eerie sound of a Bornean siamang gibbon. The effect is transcendent, beyond anything achieved by Messiaen or other composers who have attempted to transcribe animal song. As Blackford said when I spoke to him: "It's possible by sampling technology to have the sound of the musician wren from the Amazon, which repeats an extraordinary forty-four note cycle, or the common potoo, with its unusual pentatonic call, literally interweaving with the instruments of the orchestra so that you hear nature's original and then the so-called artistic or musical version of it." This unique admixture of human and natural sound is an entrancing introduction to biophony that should form part of every school curriculum.

The niche thesis is at the very least an aesthetic and appreciative lens through which to appreciate the sonic output of all living things except man. I love its central concept that everything listens to everything else, and thus finds a space in frequency and time to make its own unique sound. As the world-renowned primatologist Jane Goodall said, "The Great Animal Orchestra speaks to us of an ancient music to which so many of us are deaf." I passionately *want* to believe that nature listens, and I think the evidence is that, in the main, it does indeed. Perhaps it's just we who need to open our ears.

From single-celled organisms to a 200-ton blue whale, we find sound being made and listened to: mating calls, warning sounds, territorial markers, threat noises, hierarchy signals and social interactions, language and song. As the range of sounds for an animal to interpret extends from simple threat and food detection to teamwork, social organization and even conversation, hearing occupies ever more neural resources. The brain has to filter useful information from background noise, and then interpret sounds. This is *listening*, as distinct from hearing, which is why my definition of listening is "making meaning from sound."

All together now

But to listen to animal sounds, we must always keep in mind that not everyone listens like we do, which is true of other humans as well as other species. Life on Earth has myriad ways of perceiving and processing sound, much of it beyond our perception. A soundscape for a bat, dolphin, whale, elephant or grasshopper is as alien to us as would be a landscape viewed in ultraviolet light or x-rays. When you listen to natural soundscapes, you are perceiving only part of the information experienced by your dog or cat. Approaching nature with humility, we can more completely appreciate and perhaps glory in the stunning soundscapes that abound all over the planet but are being steadily eliminated by our cloth-eared destruction of habitats. Rainforests, jungles, the plains of Africa teeming with life, noisy seashores both above and below the water, arid deserts, enigmatic deep oceans, temperate forests with massive trees, the huge frozen taiga and frosty tundra, grassy steppes and Pampas, the magnificent peaks of the world's mountain ranges...all these and so many more natural soundscapes are wildly different, and every place within them has its individual voice at any given moment.

That's because soundscapes vary not only in space but also in time. A jungle at night sounds completely different from the same place in the daytime, and a rainforest is radically altered from one minute to the next when a torrential downpour raises the noise floor by 60 or 70 decibels. If the joy of the experience is not enough to persuade us to listen, there are other reasons that may be even more important.

Why biophony matters

There are at least three significant reasons for listening more closely to the sounds of life around us.

First, enhancing our husbandry with applied sound could yield obvious benefits. At the smallest scale, we know that cells vibrate at audible pitches, and it seems likely that they use this sound to moderate their behavior. We have seen in this chapter how playing sound to brewer's yeast cells stimulates the growth of cells and influences their chemistry. The interaction of cells and sound must surely be a key area for research: the possibilities for spotting and controlling disease (including cancer), or for stimulating beneficial cell growth, seem incredibly exciting.

In the plant domain, Israeli researchers who identified plants audibly complaining when drought-stressed suggest that one application of this discovery would be to set up ultrasonic monitors on fields of crops, so that unhealthy water shortage (or possibly even pest infestations) would be picked up much earlier by detecting the changing sound of the plants and sounding an alarm. Plant infestations make noise in their own right. The Natural History Museum team are listening to soil with embedded microphones, because many beetle larvae developing underground stridulate; hearing the sound would allow detection and treatment before the larvae hatch and damage is done. Also, beetle infestations in dead timber make a sound, so it would be possible to check imports using sound, and quarantine wood that contains undesirable pests. Another interesting reason for putting small microphones in trees is to spot damage by noticing a change in the way they creak or resonate in wind and sounding an alarm before a limb or even a whole tree falls in a street. Computers, Wi-Fi, solar cells and artificial intelligence (AI) make all these concepts increasingly achievable, and they may well become commonplace procedures in the future.

Secondly, sound can reveal changes in the natural world that we can't perceive with eyes alone—in particular, the often-devastating effects of our own behavior. Bernie Krause relates a perfect example of this in his writings and his TED talk. In 1988 he managed to get permission to record the natural soundscape in Lincoln Meadow at Yuba Pass, two thousand meters above sea level in Sierra County, California—just before a lumber company carried out a program of "selective logging" that their scientists assured the community would have no negative impact at all on either wildlife or aesthetics. The recording contained sounds from woodpeckers, quail, sparrows and many insects, and the sonogram confirmed a dense, disparate soundscape.

A year later, after the logging, Bernie returned, ensuring that the context was as similar as possible to the previous year's recording. At first glance, it seemed that the logging company had delivered on its promise: the site looked very much the same. But as soon as the microphone started to pick up the soundscape, it was clear that devastation had been wrought. The rich and varied soundscape was now dead: the only audible living thing was a single woodpecker. Bernie walked past the front line of trees and discovered great bare patches in the forest where all the trees had been taken. A cursory visual inspection from the meadow would see only the veneer of healthy trees, but the forest had been gutted and the wildlife had gone. It never returned. Bernie told me that he has recorded fifteen thousand identified species in five thousand hours of precious sound recordings. Why precious? Because over more than half of these sounds have now disappeared completely, never to be heard again.

The third and possibly most important reason for listening to nature is that there is so much we don't know about the living things around us. Decoding the languages or symbols used by animals, plants and cells will help us understand both them and our own relationship with them far better than we do now. This is crucial because we are part of nature, not above it, and this sound can connect us. A perfect, though harrowing, example of this is the recording of a beaver, described in Bernie's book *The Great Animal Orchestra* as "the saddest vocalization I've ever heard." At a small lake in Minnesota, game wardens for some reason dynamited a beaver dam that had been there for many years. That evening, a sound recordist captured the noises made by the surviving, probably wounded, male swimming in slow circles around the pond, mourning his dead family. It is not possible to hear this heartrending sound and disengage emotionally from our devastating impact on the animals with whom we share this planet. You can hear it for yourself in Bernie's TED talk [50].

Just a few hundred million years after the Earth cooled, geochemistry became biochemistry, and single-celled life formed, probably in geothermal vents. The speed of this transition (in terms of cosmic time) would suggest that microbial life may be relatively common in the universe, because there are probably millions of worlds in our galaxy alone that will duplicate those conditions with water and a hot core. Mars was one before it lost its atmosphere, so soil samples returning from the red planet in the early 2030s may confirm the abundance of life in the universe at last by showing that it did arise independently on another planet. This will be one of the great transformational moments in human knowledge.

However, it has taken around four billion years for our civilization to evolve. That's a third of the age of the universe, and eighty percent of our planet's lifespan: the Sun's expansion will render the Earth uninhabitable in a billion years. *Homo sapiens* has existed for only 200,000 years, so for 99.995 percent of life's existence on Earth, it was not intelligent. We have been using metal for only a few thousand years, and discovered how to generate power less than two hundred years ago. Our impact on the world as we proliferate is depressingly negative and potentially catastrophic, while it would take only one massive solar event, supervolcano or large meteor impact to put us back to the stone age, or even eradicate us. Our intelligence is more fragile than we like to pretend.

It may also be far more precious. There are possibly a trillion species on Earth and millions of types of multicellular organisms, but only one that has developed a civilization. Even with the recent realization that almost every star has planets, and the resulting calculation that there are probably 300 million planets in our galaxy in "Goldilocks zones" that may have atmospheres and water and could support life as we know it, the odds against another civilization having evolved are very long. Recent detailed work by scientists at Oxford University's Mathematical Ecology Research Group, building on Brandon Carter's seminal statistical model of life evolving in the cosmos, reaffirms that intelligent life is likely to be exceptionally rare.

For over forty years we've been avidly listening for electromagnetic signals that might indicate life elsewhere, and in all that time only two signals remain unexplained: the famous Wow! signal from 1977 which now appears to be from a star 1,800 light years away; and a more recent version in 2020 named as Breakthrough Listen Candidate 1 (BLC1), from Proxima Centauri, a red dwarf star with two planets that's just 4.24 light years away. Neither signal has ever repeated, and the scientific jury is very much out: non-alien explanations are considered massively more likely than E.T.

In the absence of evidence to the contrary, we would be wise to assume that our civilization and intelligence are indeed exceptionally rare, if not unique. This perspective might just create the humility we need to temper the arrogance and disconnection which drives our brutal abuse of our home planet, other living things, and even other human beings.

The last word is from Bernie Krause: "When we lived closer to the natural world, we sang as part of that animal chorus. We found a niche for ourselves that fitted within that chorus. We imitated rhythms that connected within that chorus. Now we have a disconnect: we are not quite as healthy, vibrant or aware of the world around us. We have to shut the f*** up!"

WHAT YOU CAN DO

Spend time in one of the sound libraries mentioned in this chapter and marvel at the richness and diversity of living sound. Great places to start are birdsong and whalesong, but the range is almost limitless. Read some of the fine books on the subject, such as Bernie Krause's *The Great Animal Orchestra*, Seth Horowitz's *The Universal Sense*, David Rothenberg's *Why Birds Sing* and *Bug Music*, or Karen Bakker's *The Sounds of Life*: you'll find these and many more listed in the Further Investigations section at the back of this book. And surround yourself with biophilic sound, ideally real, though recordings will also work: it's good for you.

*

Now it is time to deal with one of the smallest of the subsections of animal species, but the one that makes a disproportionate amount of noise: humans.

5

ANTHROPOPHONY

The wonted roar was up amidst the Woods,
And fill'd the Air with barbarous dissonance.
                                        John Milton

The sound of humanity is coruscatingly beautiful and hideously ugly: a sweet mother's lullaby and a guttural shout of rage; a Steinway piano and a chainsaw; Mozart and traffic noise; Pavarotti's "Nessun dorma" and the scream of a torture victim; the oratory of Martin Luther King and that of Adolf Hitler; the exultant roar of a football crowd celebrating its team scoring and the frenzied yells of a rioting mob; the mellifluous tinkling of wind chimes and the tuneless beeps of a row of supermarket checkout tills; courteous conversation and violent argument; Gregorian chant in Solesmes Abbey and Kiss at 136 dB at the Ottawa Bluesfest; the painstakingly careful raking of pea gravel in a Zen rock garden and the random chatter and ringing phones in an open-plan office...

Humanity is the planet's biggest noise-generator—not intentionally, but simply from just being human.

Noise on land…

Since the taming of horses and the invention of the wheel some six or seven thousand years ago, human beings have created increasing amounts of noise simply from moving about. The wealthiest citizens of Rome, the world's first city of a million people, pressured Julius Caesar to ban horses and carts—but only during the day. Presumably the bedrooms in their luxury houses were far enough away from the roads that night-noise was not a problem!

Human movement started to become seriously noisy with the arrival of the steam engine, the power source of the Industrial Revolution. By the late nineteenth century, champing "iron horses" were traversing the American continent on the Union Pacific Railroad and the Canadian Pacific, and the luxurious Orient Express was running wealthy adventurers between Paris's Gare de l'Est and Constantinople (now Istanbul). Diesel and electric engines eventually replaced steam, but trains are still noisy at speed, as anyone standing on a platform when a fast train comes through can testify—especially if you are looking the wrong way and don't see it coming. Notwithstanding the relatively serene Metro in Paris, underground trains can be even louder as all the train noise reflects from the tunnel walls, especially on the older networks like those in London or New York. In Patrick Shen's outstanding film *In Pursuit of Silence*, I am shown measuring the noise from a London tube train peaking at 107 dB—*inside* the carriage.

Many people rejoiced at the thought that the world would become quieter when motor cars arrived, with their pneumatic tires replacing the racket of horseshoes on stone. Little did they know! In the twentieth century, carbon-based fuels produced a quantum leap in transport noise by democratizing it: now, everyone could be a transport noise generator. The internal combustion engines in cars have become much quieter since the early days, but there are now 1.4 billion of them, according to live-counter.com, with three new cars being built every second, so the total amount of car noise is vast and growing, especially in countries with massive cities and a large stock of old-fashioned vehicles still on the road due to low per capita income. Growling heavy diesel vehicles aside, the engines of modern cars are very quiet, especially those of electric vehicles, which in Europe will be legally obliged to make a warning sound at low speeds as they are so hard to hear. But the problem won't go away when we are all-electric: tire noise (which is the major element of traffic noise on faster roads) remains unavoidable until someone invents antigravity.

The UK's Department for Food, Agriculture and Rural Affairs estimates that the annual social cost of urban road noise in England is between £7 and £10 billion, placing it at a similar magnitude to road accidents (£9 billion). Traffic noise in cities can average over 80 dB for long periods, peaking at well over 100 dB, with devastating effects on not just human health; we have already met cockney ducks and shouting nightingales—rare verifiable examples of the impact of urban noise on other species.

After transport, humanity's next noisiest land-based peacetime pursuit is construction. Cities must be built, and the tools and machines to do this have become much noisier over time. Heavy vehicles with reversing alarms, pile-drivers, tunneling machines, pneumatic drills and power saws can all generate over 100 dB of noise, added to which there is manual hammering, sawing and, of course, the ubiquitous loud radio in building sites. Ear defenders, known in the U.S. as hearing protection devices, are obligatory for construction workers on sites averaging 85 dB in both the U.S. and Europe, though this law is widely flouted: the U.S. National Institute for Occupational Safety and Health reports that more than half of noise-exposed construction workers don't wear hearing protection, so it's not surprising that one in four of them have a material hearing impairment. Meanwhile, anyone living or working next door to a building site knows how well noise travels, both through the air and in vibrations through buildings or sometimes the ground itself.

In cities, the effects of construction on wildlife may not be devastating, largely because there isn't so much wildlife there to start with—but construction noise often blights the countryside too, as roads, railways, industrial facilities or new towns are built, with serious consequences for flora and fauna both during and after the actual construction.

Industry is another major global noisemaker, from mining and quarrying to metals, chemicals, manufacturing, energy, food and agriculture. Many industries use heavy machinery in workplaces that are just as noisy as building sites: a 2004 study, for example, found that over twenty percent of workers in corn mills and sawmills showed evidence of noise-induced hearing loss. Industrial noise inside a building affects mainly the people working there, as wildlife has probably long since vacated the area—but outdoor industries like mining, logging, farming and fishing create noise pollution along with their other environmental impacts. At the extreme, practices like fishing with explosives, which is still tragically common in Southeast Asia, have devastating effects on local ecosystems for pathetically little return: most of the dead fish sink to the bottom.

When visiting my clients at Armstrong Ceilings in Lancaster, Pennsylvania, I always avoid the freeways on the drive from Philadelphia Airport and take the minor roads, which wend their way through some lovely farmland owned by the large Amish community living in that area. They eschew all machinery and farm their land using just the strength of their own bodies and those of their horses. It's a peaceful place, though the toil must be unremitting. Working that way maintains a deep physical connection with, and respect for, the land that farmers have had for around twelve thousand years, as well as preserving the natural soundscape. Machinery may be far more efficient, but it changes the relationship between human and nature into one based on maximum exploitation, even abuse, with plenty of noise occurring as a side effect, with its own damaging consequences.

Add all this up, and it's becoming difficult for land animals to find the auditory environments they have evolved to thrive in. Noise harms their hearing, makes their habitats inhospitable, damages their health by scaring or intimidating them, and eliminates them by masking danger sounds and interfering with their hunting for food or their critical breeding-related communication.

…and at sea

As we traversed the oceans, sail gave way to steam then diesel and gas turbine engines, and our noise grew exponentially. The cavitation of huge propellers is loud and it's focused in the lower frequencies, which happen to be the auditory bands used by many whales. There are now over sixty thousand commercial vessels moving around the seas at any one time, mostly in the northern hemisphere, while recreational vessels and ferries dominate the soundscape closer inshore. As the Arctic ice retreats, even that rare quiet refuge is disappearing fast: Russia is ramping up traffic on the Northern Sea Route to transport gas, oil and other products, while Canada's Arctic waters saw traffic triple in the fifteen years to 2015.

Sound travels much further and faster in water, so the effect of our global shipping noise is that the oceans are now an auditory fog where cetaceans and other animals relying on sound for navigation, communication, mating and hunting are virtually deafened. An extensive body of research has analyzed and proved beyond doubt the plethora of negative effects that anthropogenic noise is having on marine life.

A series of mass whale strandings and beachings starting in the 1980s points at a more potent sonic impact than even propeller noise. Beak whales are the free-diving champions of the animal kingdom, descending to depths of 3,000 meters and staying underwater for hours on a single lungful of air. Perhaps this deep-diving ability, involving that storing of oxygen throughout their bodies, is why hundreds of them have been washed up dead in multiple events on the Eastern seaboard of the USA, the Caribbean and the Western Isles of Scotland. The likely culprit is active sonar, a massively powerful sound that seems to drive these whales into panic, causing them to come to the surface too fast and die from the bends.

Active sonar is not the only aggressive sea noise we create. Oil and gas exploration involves repeatedly firing underwater seismic air guns for months at a time, often in tuned arrays of multiple guns firing together. These guns reveal subsurface geological features by rapidly releasing compressed air in a bubble to create a deafening sound at frequencies of around 10 to 50 Hz that penetrate the sea floor, going as far as thirty kilometers under the surface. The reflections of this sound are recorded by hydrophones, then computers analyze the data to create a picture of the underlying geology. The noise can peak at over 200 dB using the underwater scale, which equates to almost 140 dB in the more familiar airborne scale—well past the threshold of human pain and instant hearing damage—and it can travel hundreds of kilometers in sea water. The only solid research on the impact of these guns on marine wildlife relates to potential hearing loss, which appears to be mild in fish (unsurprisingly since most don't hear well) but more serious in cetaceans. The only other evidence is that whales and dolphins seem to alter their behavior within audible range of the guns firing, diving less or moving away. I'm not aware of any research assessing the effects of masking, where this noise stops marine life from hearing other important sounds. The animals have no mechanism for telling us of this, or of the degree of any pain or permanent damage inflicted, so much more research is needed. Meanwhile, the firing continues.

Added to ships and mining is a new source of noise: the construction and operation of thousands of offshore wind turbines. These, too, require seismic surveys (albeit with quieter systems than the air guns used by oil and gas prospectors) and their construction involves the noise of driving piles into the seabed, along with sound from service vessels during construction and throughout operation; decommissioning at end of life will also be a noisy business. However, once built, operational noise seems unproblematic: UK and Danish monitoring has shown the regular presence of harbor porpoises and seals around wind farms. There are also benefits to put in the balance: fishing vessels do not go near offshore wind farms, so they form a safe habitat, while their foundations create a sort of artificial reef that seems to encourage new life to move in and thrive.

There is hope for reducing our sea noise. New technologies can reduce the noise from propeller cavitation and engines, though retrofitting systems to thousands of existing ships is probably a pipe dream—however, bubble curtains that can contain the sound of seismic guns are being increasingly deployed. The International Maritime Organization is working to update its outdated guidelines for the reduction of underwater noise while, despite the continuing lack of any international regulation, there do exist some local schemes: for example, ships arriving in Vancouver pay reduced harbor fees if they meet defined noise and pollution criteria. The year 2022 saw the fiftieth anniversary of the U.S. National Oceanic and Atmospheric Administration's National Marine Sanctuary System, which protects a network of underwater parks encompassing over 1.6 million square kilometers of marine and Great Lakes waters—though, for perspective, seawater alone covers over 361 million square kilometers of the Earth's surface, so this literally is a drop in the ocean. The 2023 UN High Seas Treaty took a decade to agree, and could designate a third of the world's oceans as protected areas by 2030, with limitations on shipping lanes as well as mining and drilling—but only if it ever actually comes into force: it needs sixty countries to sign up and pass the legislation in their own countries, which is likely to be a very long haul.

WHAT YOU CAN DO

Sea noise is even less likely to exercise the public than urban noise, simply because it's so far removed from our daily experience. But we depend on the seas for food, and marine animals are a vital part of the global ecosystem. Plastic waste, overfishing and climate change are bad enough without deafening the many animals that rely on their hearing to survive. This makes it vital that the evidence of our crippling impact gets communicated, especially to children.

One way of doing this is to have them watch *Sonic Sea* [51], an Emmy-winning documentary about the impact of industrial and military ocean noise on whales and other marine life. It tells the story of a former U.S. Navy officer who solved a tragic mystery and changed the way we understand our impact on the ocean. The film is narrated by Rachel McAdams and features Sting, plus renowned ocean experts Sylvia Earle, Paul Spong, Christopher Clark and Jean-Michel Cousteau. It is available free to schools, with an accompanying study guide. To request a copy for your alma mater or your children's school, email allie@rocofilms.com.

...and in the air

The loudest transport-related noise ever made has come from the vast engines of space rockets. Once a rare occurrence, the increasing demand to put satellites into orbit means that rocket launch sites are becoming profitable enterprises and are sprouting worldwide, usually in remote areas. Not far from my home in remote Orkney, the

SaxaVord Spaceport on the Lamba Ness peninsula in the Shetland Islands was the site of Scotland's first space launch.

Rocket noise can be literally earthshaking. The new NASA Moon vehicle, the Space Launch System, is a huge rocket, fifteen percent more powerful than the iconic Saturn V that took the Apollo astronauts to the Moon. A team from Brigham Young University in Utah measured the noise output from the first SLS launch in Florida on 16 November 2022 at over 120 dB at a distance of several kilometers. At the launchpad it was probably over 200 dB. With the growth of private "launch providers" like SpaceX, and the concomitant spread of rocket launch sites, many people are now working hard to estimate the effects on local wildlife. Launches may be relatively infrequent and short in duration, but when our local launch site is fully operational, its intensity will certainly be unlike anything the seabirds of northern Scotland have ever experienced before.

When the Wright brothers touched down after their epochal flight on 17 December 1903, they could never have envisaged its long-term impact on the planet. In 2019, the last full year before the Covid pandemic, just under 4.5 billion passengers and 57.6 million tons of freight traveled by air on 38.3 million flights; by 2022, traffic had rebounded to similar levels as FlightRadar24 tracked around 200,000 flights a day, of which over half were commercial.

Planes are the noisiest regular transport of all. At low altitude in take-off and landing, aircraft noise blights the lives and well-being of millions of people living close to airports. Francesca Dominici, Professor of Biostatistics and Associate Dean of Information Technology at Harvard School of Public Health, led one 2013 U.S. study and commented on the findings: "Living close to an airport, and therefore being exposed to aircraft noise, can adversely affect your cardiovascular health, even beyond exposure to air pollution and traffic noise." An earlier UK study found that the risk of hospital admission for stroke or coronary illness was twenty to thirty percent higher in the noisiest areas than in the quietest, while a 2023 study by the University Medical Center, Mainz, found that aircraft noise exposure before a heart attack substantially amplifies subsequent cardiovascular inflammation and aggravates ischemic heart failure. Cardiologist Thomas Münzel commented: "There is no doubt any more that transportation noise must be considered an important cardiovascular risk factor, comparable to hypercholesterolemia, hypertension, smoking, and diabetes mellitus." Health is not the only cost of low-level aircraft noise: children in schools or homes close to airports fall behind their peers in quieter locations—for example, a 5 dB increase in aircraft-noise exposure was associated with a two-month delay in reading age in a UK study. Perhaps the biggest impact is on sleep, with all the consequent health damage that is now well documented.

Even at altitude, jet noise disrupts the pristine tranquility of some of the remotest places on Earth. The shrinking ice sheet of central Greenland is overflown hundreds of times a day by planes traveling between Europe and the USA, while flights from the USA to the Far East regularly disturb the great wildernesses of northern Canada and Alaska. The deepest rainforests and jungles and the remotest deserts suffer the same sonic gate-crashing. Finding places where air traffic does not intrude requires traveling to a dwindling number of remote places: Antarctica's soundscape is never polluted by commercial airliners, while the Arctic Circle wildernesses of Siberia and northern Scandinavia remain largely untouched, along with some parts of Africa and the Australian outback.

In many natural beauty spots closer to human habitation, even when jets are not imposing their noise, other forms of anthropophony often do: chainsaws, motorbikes, cars, leaf blowers and lawnmowers can pollute soundscapes for kilometers in all directions, even in national parks. Sadly, people often associate the noise output of such devices with power, so trying to sell quiet versions can be challenging. Anthropophony is increasingly pervasive, as nature sound recordist Martyn Stewart told *Condé Nast Traveller* in 2012: "In 1968, it took only three to four hours to get one hour of pristine natural sound. Today, it takes up to two thousand hours."

For all its terrible consequences, Covid-19 created a unique pause in humanity's remorseless noisemaking. In spring 2020, global lockdowns left capital cities as ghost towns, flights and road traffic almost disappeared, and some strange things happened, as *Newsweek* reported: "Whales migrated closer to shore due to the lack of ship noise interfering with their sound location, wildlife was free to roam—a puma sauntered through Chile's deserted capital city; coyotes were spotted near San Francisco's Golden Gate Bridge—and in New York City, residents heard more birds than traffic." Some found it intimidating; others discovered something that had been missing their whole lives.

Human beings are inveterately noisy, especially in large groups. As we've seen in the previous chapter, we're certainly not the only noisemakers on Earth, but dense human population inevitably means noise, and as the population grows, the noise level follows. R. Murray Schafer made a rough estimate that urban noise is increasing by half a decibel every year, which would mean a city becomes twice as loud every twenty years. That's for a single city, but both the number of cities and the proportion of humanity living in them are growing too: according to the UN, in 1960 twice as many people lived in rural areas as in cities, but by 2020 the ratio had shifted dramatically and well over half of humanity was living in urban areas. In 1950 only New York and Tokyo were "megacities," boasting a ten million population; now, there are over thirty megacities, and Tokyo's population will soon reach forty million. We are getting noisier, made starkly obvious by temporary pauses like the shut-down of air traffic after the eruption of the Icelandic volcano Eyjafjallajökull in 2010, or the blanket of silence that fell over most cities during Covid lockdown.

From the Industrial Revolution onwards, mechanical and then electro-mechanical noise have taken this form of pollution to a whole new level. When I spoke with Mike Goldsmith, an acoustician turned best-selling science writer and author of the book *Discord: The Story of Noise*, he made an important qualitative point: "Suddenly there was a new type of noise, a kind of background hum... You can't isolate any particular sources of it, and this makes that type of noise particularly disturbing, because people have evolved to use their hearing as a way of identifying sounds, particularly prey or predators. When you get to the industrial period in cities, it's no longer possible to identify specific sources. This can be quite disturbing: the ear is constantly trying to isolate the elements of this background hum and to locate it in space and identify the source, and it just can't be done."

Understanding noise and its impacts

Not all noise is bad. Bart Kosko, Professor of Electrical Engineering at the University of Southern California, wrote a fascinating book simply called *Noise*. When I spoke with him, he told me that there are many benefits from introducing small amounts of noise into all sorts of systems, including our brains. He explained: "It's called stochastic resonance... If you get it just right, you can actually improve how certain systems behave, but only if they're nonlinear. We have overwhelming evidence, mathematically and otherwise, that neural-type systems benefit from noise. Again, they can be overwhelmed, but a little bit of noise can energize it just right and help you detect a very faint signal... a little bit of background noise can make you smarter or keep you more alert. Too much is too much stress, but a little bit of it, at least in my experience—and the neural models support this—ought to keep you more on your toes."

There are various colors of noise: from white and pink, to brown and black. These are arbitrary names, and they describe the frequency profile of noise. White noise [52] has equal energy across all frequencies and sounds like hiss. Pink noise [53] is profiled to match the sensitivity of human hearing, so it sounds more even to us and is often used in masking; some people like it when working, probably enjoying the stochastic resonance Bart Kosko describes.

Although a little noise can be helpful, Kosko cautions: "Mammals evolved in a world of low-intensity signals, a low-decibel world. And yet the world around us is getting louder... There are just more people, more cars, there's more talking—but our hormone systems, our endocrine systems, are adapted to a very quiet environment. And that's something very difficult to balance, because too much noise means more stress, and that correlates with lots of bad health things like high blood pressure, risk of heart attack... stroke."

The American Academy of Nursing perfectly summarized the impacts of noise on health in a 2016 position statement: "It has been well documented that noise exposure contributes to hearing loss, tinnitus, heart disease, stroke, anxiety, stress, depression, learning difficulties, job performance, sleep disorders and reduced cognitive abilities. Noise is more than an annoyance; it is a public health hazard."

According to the World Health Organization, environmental noise contributes to 48,000 cases of ischemic heart disease a year, as well as causing 12,000 premature deaths annually. The most common casualty of noise is sleep: eight million Europeans are having their sleep wrecked night after night by traffic noise that far exceeds the WHO's recommended levels. Sleep deprivation and associated chronic noise-related stress have myriad consequences, none of them good: the European Union Environmental Noise Guidelines estimate that "One million healthy years of life are lost every year from traffic-related environmental noise in Western Europe." The economic burden is vast, including the costs of healthcare for conditions such as depression, anxiety and many physical illnesses that result from sleep deprivation—plus conflict and social disturbances, accidents, lost productivity and more. The EU estimate is that traffic noise alone costs at least €40 billion every year as a result of such effects.

Sadly, even in the EU, where there has been a program of detailed noise mapping all over the continent for decades, few governments have taken any effective action to control the problem. In the Global South there has so far been little official noise mapping; the pursuit of economic growth has taken priority over any concerns about noise and its effects—especially in countries with oligarchic governments or high levels of corruption.

I remember well the noise in Mumbai, where driving involves almost constant use of the horn, and myriad aging scooters and motorbikes have holes in their exhausts, or broken mufflers:

cacophony does not begin to describe it, especially on festival days, when drumming and firecrackers overlay the rest of the noise. It's not surprising that multiple studies have revealed high levels of hearing loss among Indian traffic police.

There are a few brave souls pushing back in India, but they encounter strong resistance from religious and commercial groups who defend their need for noise. Nevertheless, the environmentalists have achieved some success, with the establishment of no-honking days in Mumbai and Delhi, alongside dozens of "silence zones"; also, the Indian government's National Environmental Engineering Research Institute has undertaken some noise-mapping studies, and some police are now equipped with noise meters. But, so far, progress has been modest; it will take a strong national educational program to change the tolerance of, and even positive associations with, noise, by uncovering its effects on health, well-being and the economy.

There is some hope that people are becoming more conscious of the problem: in a 2017 quality of life survey of 37,000 Europeans, a third reported they were having problems with noise. An earlier survey showed that eighty percent of Europeans believe that noise affects their health. But despite all this, noise is not a political issue anywhere I know of. I've never heard a politician say, "Vote for me, I'll make it quieter!" Perhaps the issue is the frog-in-a-pan syndrome: the noise level ratchets up a little every year and we end up standing on street corners or sitting in restaurants, bellowing at each other and thinking it's normal. There are a few organizations lobbying for a quieter world, such as the UK's Noise Abatement Society, which was established in 1959 and managed to persuade Parliament to pass the Noise Abatement Act in 1960, making excessive noise a criminal offense. There are also a few people around the world bucking the trend by creating quiet spaces in cities, such as the High Line in Manhattan (ironically now among the most crowded spaces in the city) or urban quiet parks like London's Hampstead Heath or Barcelona's Parque del Montnegre y el Corredor.

Nevertheless, human population density and noise nuisance are always highly correlated, so noise will inevitably continue to be a growing problem. In 2022 the United Nations Environment Programme published a report called *Frontiers 2022: Noise, Blazes and Mismatches—Emerging Issues of Environmental Concern*. It includes a table of the noisiest cities in the world. London tops the chart for Europe at 86 dB average daytime noise level, and New York is the loudest North American city at 95 dB—but both pale by comparison with the South Asian cities of Islamabad, Moradabad and Dhaka at 105, 114 and a staggering 119 dB respectively. That's rock-concert loud [54].

Here's a table of the world's twenty noisiest cities according to the UN, with one change: they list Kupondole in Nepal, but this is a small district within the city of Lalitpur, so I've used that instead. The cities are ranked by dB (LAeq), which means decibels, A-weighted to represent human hearing, with each reading being an average over a period of time to give a fair representation that's not distorted by short-duration peaks. I have added the population data (for the actual city in each case, not the metro or urban area) and then a calculation of decibels per million population. On that measure, the world's noisiest places per head are Puerto Vallarta (Mexico) and the aforementioned Lalitpur (Nepal), with dishonorable mentions to Moradabad (India), Rajshahi (Bangladesh) and Islamabad (Pakistan). Remember, a rise of ten decibels is a doubling of noise: so, for example, New York is twice as loud as London.

	City	Country	dB (LAeq)	Population (m)	dB/m
1	Dhaka	Bangladesh	119	10.3	11.6
2	Moradabad	India	114	0.9	126.7
3	Islamabad	Pakistan	105	1.0	105.0
4	Rajshahi	Bangladesh	103	0.8	128.8
5	Ibadan	Nigeria	101	3.6	28.1
6=	Algiers	Algeria	100	4.5	22.2
6=	Lalitpur	Nepal	100	0.3	333.3
8	Bangkok	Thailand	99	10.5	9.4
9	New York	USA	95	8.5	11.2
10	Damascus	Syria	94	2.5	37.6
11	Manila	Philippines	92	1.8	51.1
12=	Hong Kong	China	89	7.3	12.2
12=	Asansol	India	89	1.2	74.2
12=	Kolkata	India	89	4.5	19.8
12=	Karachi	Pakistan	89	14.9	6.0
16	Cairo	Egypt	88	10.1	8.7
17	London	UK	86	8.8	9.8
18=	Puerto Vallarta	Mexico	85	0.2	425.0
18=	Kuala Lumpur	Malaysia	85	2.0	42.5
18=	Auckland	New Zealand	85	1.4	60.7

Though there is scant information about the effects of urban noise on other species, the studies that do exist suggest that it affects them just as profoundly as it does us, largely through an impact known as acoustic masking. As we've seen, animals, especially birds, need to hear in order to establish or defend territories and, even more crucially, to reproduce: in many bird species, females assess male quality through listening to song, and may be less invested in the resulting offspring if not initially galvanized by full bandwidth reception of the mating song. Birds in areas with noisy traffic have fewer offspring, especially those species that sing at lower frequencies.

Faced with disruptive human noise, animals will either move away or try to alter their communication: they may change their timing, frequencies, volume or style of calling in order to avoid the noise, or to compete more successfully with it. In Sheffield, in the north of England, where urban noise doubled in the decade to 2001, robins have adapted and now sing more at night, while in Bogotá, Colombia, rufous-collared sparrows start their dawn chorus earlier in the morning for the same reason. Such avoidance tactics are not cures but symptoms of the impact of our noise. Other evidence suggests that we are just scratching the surface of understanding the impact of our noise on other species: for example, a 2019 University of Denver study found that traffic noise caused field crickets to mature later and so have a reduced adult (reproductive) lifespan.

Notwithstanding the loudness of transport and construction in cities, most noise complaints received by the police or local government are not about these things, but about other people's behavior. Late night parties, loud music, antisocial DIY, violent arguments and drunken street noise are some common strands. As population density rises, social behavior becomes a hotter issue, although this is probably nothing exclusively modern: no doubt the neolithic inhabitants of the densely packed houses in the extraordinarily well-preserved ancient settlement of Skara Brae here in Orkney had their own heated disputes about neighbor noise five thousand years ago.

The UK's Chartered Institute of Environmental Health carries out national noise surveys, drawing together data about noise complaints from local authorities across the country. It records that eighty percent of UK residents report being exposed to noise in their homes, with low-income households the most vulnerable. The damage of noise is inversely related to wealth, making it an additional burden for the less well-off in every society, damaging both well-being and quality of life.

The sound of war

For as long as humans have fought one another, conflict has been noisy. Arguments are loud, especially if they become violent, and noisy places are often contributing factors when conflict escalates into violence. Aggressive chants like the war dances of the Polynesians—loud rhythms made by the slamming of swords or spears on shields—and the wordless roar of a charge have long been used to intimidate the enemy in battle. Since the invention of explosives, war has been the noisiest widespread human activity by far. Gunfire can exceed 150 dB in sound energy, and explosions from artillery, missiles or bombs are even louder; these sounds are so extreme that countless soldiers (and civilians) have suffered severely damaged hearing. Today's elite military have access to systems such as the U.S. Army's Tactical Communications and Protective System made by Danish specialist INVISIO, which also supplies the UK and many other armies. These are in-ear devices that will attenuate loud impulse noise to within safe limits and yet allow radio communication (using bone conduction microphones) and amplify quiet sounds like footsteps with full localization.

But even these devices would struggle in the event of the loudest war sound of all. Only two nuclear bombs have ever been used in anger, and by today's standards they were small—but they still killed some 200,000 people in Hiroshima and Nagasaki (nobody knows exactly how many). The largest nuclear device ever detonated, the Soviet Union's Tsar Bomba test on the Arctic Ocean islands of Novaya

Zemlya in 1961, was over two thousand times more powerful, shattering glass windows 780 kilometers away and creating a zone of total destruction larger than the city of Paris. The sound energy of nuclear explosions like this at source is probably well over 200 dB.

Along with the intense heat and radiation, nuclear explosions create a lethal shockwave, traveling at hundreds of times the speed of sound and pulverizing buildings and people for some distance until it degrades into sound. The pressure wave of the Tsar Bomba circled the globe three times, matching the Krakatoa eruption. Sadly, humanity has managed to manufacture something that comes close to the destructive power of nature's most cataclysmic forces—meteors, earthquakes and volcanoes.

Sound itself has also been developed as a weapon. At my first ever TED conference in 2004, I saw American inventor Woody Norris demonstrate an ingenious directional device that projected two ultrasonic waves, with the difference between them varying constantly to create an audible sound that appeared to be originating right by your ears—but only when the device was pointing directly at you. He named it HyperSonic Sound and won the 2005 Lemelson–MIT Prize for the invention. Norris was already working with the U.S. military to weaponize the invention in a much more powerful form. In November 2005, Somali pirates armed with machine guns and rocket-propelled grenades attacked the cruise ship MV *Seabourn Spirit*, expecting easy pickings. But the ship was equipped with the big brother of the device I saw at TED, a sound cannon called the Long Range Acoustic Device (LRAD), which is powerful enough to project sound long distances at 150 dB, far beyond the human pain threshold. It's not possible to hold a gun or control a motorboat with both your hands over your ears, so in the face of the terrible noise the pirates retreated. However, in 2008, pirates mounted a similar attack on the chemical tanker MV *Biscaglia*. On this occasion, when security fired up their sonic weapon, the invading crew just shrugged it off. The security team later told the British newspaper the *Daily Mirror*: "We thought it would make the pirates back off, but they just laughed. It was a total waste of time."

While the LRAD, if under careful control, may have some potential benefits for emergency communication—for example, giving targeted guidance to panicking people in a noisy emergency—its use in conflict is dangerous because, while non-lethal, when used at full power it can cause permanent hearing damage from a distance of over 300 meters. The thought of LRADs being used for crowd control by untrained or uncaring security forces in repressive regimes (or even in democratic ones) is deeply disturbing. We make enough noise on this planet without adding that to the mix.

Equally disturbing are the suggestions that Russia has been using a sonic weapon to harm U.S. and Canadian government and military personnel. A series of "anomalous health incidents" became notorious as Havana syndrome, named after the first such occurrences in Cuba in 2016. The victims have reported symptoms including confusion, memory loss, ringing in the ears, pain, nausea and even permanent hearing loss; often they've reported hearing strange sounds. No cause has been proven, but a March 2024 report by Russian investigative group The Insider with U.S. show *60 Minutes* and German newspaper *Der Spiegel* pointed the finger at the notorious Unit 29155 of Russia's GRU military intelligence service. If the accusation is true, the nature of the weapon is unknown, though any sonic device targeting specific individuals would have to use high-frequency sound tightly focused in a beam, as with the LRAD device.

Whatever the truth behind Havana syndrome, weaponizing sound is a disturbing concept.

Technology to the rescue?

Technology may be the driver of most human noise pollution of the natural world, but it can also help in the fight against it. Michel André is a professor at the Polytechnic University of Catalonia and director of the Barcelona-based Laboratory of Applied Bioacoustics—and a world leader in applying technology to balance the interests of industry with the good environmental status of the planet. He won a 2002 Rolex Award for Enterprise for his pioneering acoustic Whale Anti Collision System, which he then developed into a global passive acoustic monitoring project called Listen to the Deep-Ocean Environment. It started as a network of deep-sea microphones that listen continuously to the ocean soundscape then, through collaborations with conservationists such as the team of the late José Márcio Ayres (also a Rolex Laureate), it expanded to include land-based microphones in the Amazon rainforest, the Arctic, South America, Asia and Africa. It's impossible to monitor all these streams of audio manually, so André's team developed AI-based software that processes and analyzes the real-time data so that scientists can extract vital information about human and natural sounds, including deepening our understanding of cetaceans. Michel André has gone on to create the Sense of Silence Foundation, funded by Rolex and Monaco's Prince Albert II Foundation among many others, which deploys the technology he has developed to set up passive listening stations specifically focused on the world's most vulnerable habitats.

Over the years, I've had various conversations with people seeking to use passive listening systems with smart software for many purposes—for example, to alert police to violence or crime by intelligently spotting its sound (which would be far faster than waiting for an emergency call); or to sound an instant alarm in the event of illegal logging or poaching, in much the same way that the world's seismologists and vulcanologists listen for early warnings of major events. In populated places, there are major privacy issues with such systems, but in the wild, no such problems arise. It's inspiring to come across someone using intelligent passive listening to protect the planet's flora and fauna and to better understand (and hopefully mitigate) the noise our species inflicts on others.

Michel André's work takes this idea of using sound to track humanity's impact on the global ecosystem to a whole new level that's become possible only now thanks to AI. Meanwhile, a more old-fashioned approach using intentional sound may help to regenerate coral reefs, important marine habitats that are being devastated by global warming, overfishing and pollution so badly that the world has lost half its coral reefs since the 1950s. Scientists from the Woods Hole Oceanographic Institution in Massachusetts found that playing recordings of healthy reef sound almost doubled the number of coral larvae settling, while experiments in Australia's Great Barrier Reef have established that the same sounds also attract young fish as they return from deeper waters where they mature after birth. While these methods won't reverse the damage from warmer seas and pollution, they do at least give hope that coral will not be completely lost.

Building sound

So far, we've looked mainly at the sound that humanity produces unconsciously or with negative impact. But, of course, there is a whole other class of sounds that we make intentionally. Throughout human history, when groups or communities come together for social gatherings, sound has played a vital role. When I visit the neolithic stone circle and henge known as the Ring of Brodgar on Orkney, I find it impossible not to imagine the sound that was made when using it. The stones date back to 3100 BCE, six hundred years earlier than Stonehenge. Orkney people were then living in small communities in tiny dry-stone-walled houses and they had little spare time—with no machines, simply fishing, farming cattle, growing rye, and staying warm and dry in the Orcadian weather would have been almost a full-time job. Nevertheless, someone persuaded these people to come together to quarry dozens of enormous stones, drag them over rough ground, probably using just seaweed to lubricate their progress, and erect them in these dramatic monuments. We don't know for sure what sounds would have been made in the completed structure but, from many other ancient constructions, we can infer that sound played a much more important role in their creation than it does in today's buildings.

In a fascinating paper published in 1999, Reading University's Aaron Watson and Dave Keating established that two ancient structures (a recumbent stone circle near Aberdeen and a passage-grave chamber in Caithness) were both "ideal environments for producing dramatic sound effects" due to their resonant properties. The authors also calculated that the configurations of various passage-graves in

Scotland, Ireland and Wales were likely to generate Helmholtz resonance—a special form of resonance, associated with the characteristic shape that all these chambered cairns share (a large inner chamber with a narrow entrance passageway), which can powerfully amplify certain frequencies. In all the chambers they analyzed, Watson and Keating found that the resonant frequencies were infrasonic, ranging from 2 to 7 Hz. These frequencies are too low to be sung, but they can certainly be drummed. The authors ran some tests in the Camster Round tomb in Caithness, and their volunteers reported a range of unfamiliar sensations occurring only when the resonant frequency was being drummed out. Many of the neolithic structures here in Orkney are described in guidebooks simply as burial chambers, but when I visit them I feel convinced they had other functions: for example, allowing ancient shamans to achieve altered states by chanting or drumming in their resonant acoustics. Our ancestors' relationship with sound must have been very different to ours. After all, without the scientific explanation for thunder and lightning or sunrise, their capacity for believing in the miraculous, mysterious and unseen must have been much greater.

Ten miles south from the Ring of Brodgar, on the island of Hoy, inland from the huge sea cliffs of Rackwick at the center of a peatland valley is a mysterious slab of stone known as the Dwarfie Stane. It's a massive block of dense Devonian old red sandstone, dropped by a retreating glacier at the end of the last Ice Age. It is unique in northern Europe in having been painstakingly hollowed out to form a chamber with two separate cells, with perfectly smooth walls and floors; one of them has a pillow formation at one end. The entrance stone, which would plug the doorway, lies in front. The hollowing out of this enormous block was done by people some five thousand years ago using simple tools made of stone and antler—an almost inconceivable feat of endurance, determination and patience. No bones or other contents have ever been recorded as found in the stone. It is nevertheless dubbed a rock-cut tomb, though it would be the only one in northern Europe and bears little resemblance to the ornate necropolises of later Mediterranean cultures like Egypt.

I have sat in the Dwarfie Stane and experienced its resonant frequencies, imagining how it might have been five thousand years ago for a shaman, chanting in tune with the resonances of the space in pitch darkness, wrapped in the invisible and dissolving the walls of whatever spirit world he or she believed in.

Researchers have identified another much larger structure of similar age that appears to have been specifically built for this purpose. The Hypogeum of Ħal Saflieni in Malta is a UNESCO World Heritage Site, believed to be the oldest underground temple in the world. Parts of the structure have an extraordinary reverberation that lasts up to thirteen seconds, and in a 2015 paper, archaeoacoustic researchers Paolo Debertolis, Fernando Coimbra and Linda Eneix found that its Oracle Room exhibits an overwhelming double resonance at 70 Hz and 114 Hz which creates the sensation that sound is coming from everywhere. Those who have experienced the sound have reported feeling it intensely in every tissue, including a buzzing in the ears. Resonance at 70 to 130 Hz turns out to be a very common feature of megalithic structures; for example, the famous passage-grave at Newgrange in Ireland resonates strongly at 110 Hz. The question is, why?

One prosaic reason is that this is in the lower range of a typical male voice: basses start at about 80 Hz, and baritones at about 100 Hz. If ancient shamans were mainly male, these chambers would powerfully enhance their singing or chanting. But neurological research suggests the frequencies may also have had a potent effect on the listeners: a 2008 paper by a team from UCLA led by Ian Cook reported that in tests a frequency of 110 Hz caused listeners' left temporal region to become much less active, and the right hemisphere to dominate neural activity. The left hemisphere is the principal center for speech and rational thinking, while the right is the dominant side for emotion, sound and music perception and dreaming. Further research by Paolo Debertolis found that people vary in their "frequency of activation" of the right hemisphere, but the range is always between 70 Hz and 130 Hz.

Of course, the most famous ancient structure of this sort is Stonehenge. Professor Trevor Cox, head of Acoustics Research at Manchester's Salford University, collaborated with fellow acousticians Rupert Till and Bruno Fazenda to analyze the acoustics of a perfect replica of Stonehenge in Maryhill, Washington State, chosen because the original is missing many stones now. Trevor told me what they found: "It's like the difference between talking outside and going into a room—so not necessarily specific resonances, but more the fact that you could just hold a conversation. You don't have to be standing facing people: you could be behind the stones and you could still be heard."

However, Cox is skeptical about the tuned resonance of ancient buildings: in his book *Sonic Wonderland*, he points out that any modern bathroom has similar acoustic properties. But neolithic people didn't have bathrooms, so the spaces they created were the *first* built structures to resonate at these frequencies. It's the bathrooms that are the coincidence, not the ancient structures; the locations of cave paintings show that these people were sensitive to acoustics, and even if the first chambers they built were accidentally resonant, they would have noticed, used the sound, and then replicated it intentionally in further structures.

In 1993 Linda Eneix, author of the book *Listening for Ancient Gods*, founded the OTS Foundation for Neolithic Studies. It has since held three international conferences bringing together experts in archeoacoustics to discuss the role of sound in ancient spaces, including the thesis that ancient builders designed spaces specifically for their sonic character.

It's a fascinating area of study. Although much more research is needed, it's not unreasonable to suggest that our ancestors deliberately designed these structures so that they could experience carefully tuned chanting or drumming in order to move into altered states of consciousness. I strongly suspect that these spaces were designed with and for the ears just as much as the eyes—a focus and skill that has been lost in the millennia since.

If the acoustics of later religious buildings expressed or enhanced a connection with the divine, it was largely accidental. Christian architecture developed pragmatically, initially adapting pre-existing secular basilicas (which laid the foundations for the typical design elements of churches) before evolving into vast Gothic cathedrals with their reverberation times of up to ten seconds. Such colossal acoustics were a consequence of ocular design founded in geometry and arithmetic, designed to express the grandeur of God—rather than being the driving force behind these structures. Intense reverberation may nevertheless evoke awe and a sense of the transcendent, like in the Hagia Sophia, the former cathedral and mosque in Istanbul—or in the much smaller, cylindrical baptistery of St. John in Pisa, Italy, with its geometrical proportions creating an extraordinary twelve-second reverberation time. The acoustics of Christian cathedrals did however spawn a specific musical form to exploit their extended echoes—plainsong.

As time passed, appreciation of the power of acoustics in architecture largely disappeared, except in a few structures: spaces for education or the performance of music or drama demanded careful acoustic design, as the alternative name "auditoria" ("places for hearing") suggests—though even in these buildings, that care was not always forthcoming. Many older concert halls were built in what one could call the acoustic dark ages, after any ancient sensitivity to sound had died but before the modern science and art of architectural acoustics arose with the path-breaking work in the last years of the nineteenth century of Wallace Clement Sabine, who created the concept and formal definition of reverberation time and worked on the design of the Boston Symphony Hall. Sabine identified that the problem in many spaces was inappropriate, usually excessive, reverberation time. He suggested maxima of around two seconds for a concert hall and less than one second for a lecture hall, and established how to achieve these targets by using absorbent materials. Before this, even for specialist spaces where sound was the whole point, architects had only gut instinct to use for acoustics, and this was often trumped by visual priorities.

Thus, when the Prince of Wales opened London's enormous dome-shaped Royal Albert Hall in 1871, *The Times* reported: "The address was slowly and distinctly read by His Royal Highness, but the reading was somewhat marred by an echo which seemed to be suddenly awoken from the organ or picture gallery, and repeated the words with a mocking emphasis which at another time would have been amusing." It wasn't until 1968 that acoustic mushroom shapes were suspended from the ceiling to stop sound echoing from the huge cupola. Even so, when American acoustics expert Leo Beranek ranked fifty-eight of the world's top concert halls in 2003 by the quality of their sound based on surveys of conductors, music critics and aficionados, the Royal Albert Hall came bottom.

We have probably forgotten more than we currently know about the power of architectural acoustics. For hundreds of thousands of years, humans listened to the spaces they occupied. In caves, we listened constantly for danger, and at some point noticed the powerful acoustics and started to use them. Later, people started to construct buildings, initially rudimentary and using whatever was around purely to create shelter and warmth, but as time passed and settlements grew, aesthetics came into play and specialists arose—people who understood designing and making spaces for more than simply utilitarian purposes, which became all about appearance, not experience in all the senses. I often reflect on why almost all design awards are for how things look. This lack of recognition for sound in design is a symptom of a deeper deafness in the way we fashion our world, and it applies profoundly and widely to the spaces that architects design today.

In most modern architecture, reverberation and resonances are not design tools, but problems, because they often interfere with speech intelligibility or create unpleasant acoustics. Sadly, these issues are commonplace, because architects today are almost entirely ocular. Former MIT professor and pioneer of digital audio Barry Blesser co-wrote a superb book on aural architecture entitled *Spaces Speak, Are You Listening?* Published in 2007, the book emphasized the power of sound, the importance of listening and the vital role of acoustic architecture in maintaining social cohesion. Like my 2012 TED talk

Why Architects Need to Use Their Ears [55], it was a plea to architects to start listening to the spaces they make, for the sake of the people who have to live, work, learn, heal and play in them. When I spoke to Barry, I found him very pessimistic about the uptake of architectural acoustics since his book came out. "When I wrote the book," he said, "there was more interest in it...and that interest decayed and decayed to the point where nobody gives a shit." Why is this, I asked him? He replied, "Every architect I've ever come into contact with is interested in one thing—his career. And his career gets accelerated if he wins prizes for his architectural creation, based on visual pictures (which are stripped of people) because it's a lot easier for him to present his work in a visual domain than to say, 'You have to travel to this space and experience the sound.' Architects chose vision not for an abstract reason, but because that was the easiest vehicle to win their prizes and to get the next contract."

It's certainly true that architects largely ignore sound. In the USA, architects train for five years, but in all that time they are lucky if they spend a week learning about acoustics. And, of course, they do love hard surfaces: glass, metal, stone, wood and concrete abound in modern structures, while carpets, curtains and soft materials seem to be generally considered unfashionable. As a result, we have schoolrooms where many pupils simply can't hear their education and where teachers are shortening their lives by suffering excessive noise day after day; hospitals with noise levels twelve times the WHO recommended maximum, where patients can't sleep and recovery times are lengthened as a result; restaurants where people bellow at one another at noise levels that legally require hearing protection to be offered to workers; shopping malls with the acoustics of cathedrals and noise levels that deter shoppers; and open-plan offices where concentration for solo working is difficult, if not impossible. All these spaces create stress, fatigue and irritation, and they are all bad for well-being. In fact, they are simply not fit for purpose, purely because sound has been overlooked in their design.

Notwithstanding the continued ocular obsession of architects and Barry Blesser's pessimism, there is cause for some cautious hope that things are changing. The global schemes that promote sustainability, well-being and comfort in buildings are paying more attention to sound and acoustics. For some years, I have been a sound concept adviser to one of these, the WELL Building Standard, and the weight given to acoustics has increased markedly in the recent past; it is likely that this will soon be joined by active soundscaping for comfort, health and productivity. The first sound concept lead at WELL was Ethan Bourdeau, and he is optimistic about architects listening more attentively to their designs in the future as the evidence for the power of sound in spaces mounts up, and passionate on moving the conversation from the avoidance of negative effects (simply improving bad acoustics) to creating positive effects with designed soundscapes that support different work styles and people, including the neurodivergent. I asked him what this would involve, and he answered, "A reframing of the design process to consider acoustics as part of a trichotomy of comfort, communication and well-being [as the] three central tenants of design—rather than simply noise control for comfort, or improved audio-device performance for communication. Until now, a lot of these components have been dealt with in silos and, frankly, up until maybe fifteen years ago, health and well-being were not even a part of that conversation with respect to acoustic design. The WELL Building Standard has done quite a bit to elevate the link between health substantiation, acoustic design and, ultimately, human outcome."

There are software-based tools now that let architects "auralise" spaces before they build them. At the high end, British design and engineering firm Arup has created fourteen spaces around the world where architects or clients can hear how their designs will sound when built. I have experienced the firm's London SoundLab, with its enveloping sphere of dozens of Genelec loudspeakers. In these acoustically neutralized spaces, modern digital signal processing allows Arup's sound engineers to map the surfaces and coverings of any planned building and then accurately emulate its acoustics: choose to put a carpet on the virtual floor, and you will hear the effect instantly. With sophisticated convolution reverb software already in thousands of musicians' Macs, the inevitable trickle-down effect is happening, and there are now cloud-based auralisation apps, such as the remarkable Treble.tech, that allow architects to play with sound in their own studios. For centuries, architects have had to build things in order to find out how they feel to the occupants. Now they can check and tweak the sound before ground is broken. Perhaps, at last, they will start designing with their ears as well as their eyes.

The human voice

Throughout most of human history, the voice was the only instrument of change, inspiration and learning. The vast bulk of human speaking is lost to time, but we do have the gists and even the texts of some iconic and world-changing speeches, such as Jesus's Sermon on the Mount, Queen Elizabeth I's rousing speech at Tilbury as the Spanish armada approached, or Abraham Lincoln's timeless Gettysburg Address. And thanks to the advent of audio and video recording, we can also experience on the internet first-hand the skills of more recent orators who have powerfully influenced world events, for good or ill, such as Winston Churchill with "We shall fight them on the beaches…"; JFK, "Ask not what your country can do for you…"; Martin Luther King, "I have a dream…";—or, chillingly, Adolf Hitler and "The future belongs totally to us…"

These people may have been born naturally gifted orators, but powerful speaking is not something restricted to a gifted few outstanding leaders: it is available to everyone—though most people never do anything to learn this critical skill. I have given talks all over the world to audiences of senior managers and CEOs. I always ask for a show of hands on how many of the audience use their voice in their work, whether that's selling, leading a team, speaking to the media or speaking on a platform to an audience. Almost everyone puts their hand up. Then I ask, "How many of you have had formal vocal training?" In an audience of several hundred, perhaps half a dozen will raise their hands. This is madness!

The human voice is the instrument we all play, and yet we virtually ignore this essential life skill, neglecting to teach it effectively in school or pursue it in later life. The people who do hone it, of course, are actors and singers, and when I hear in my mind's ear the extraordinary vocal feats of Pavarotti, Aretha Franklin, Ella Fitzgerald, Robert Plant or Nusrat Fateh Ali Khan (substitute your own favorites) I mourn the widespread loss of human vocal power and glory that results from the almost universal assumption that your voice is a fixed asset given to you at birth, and just something you have to live with.

That is as untrue as the equivalent fallacy that listening is just a capability. Like listening, speaking is a *skill*, and anyone can learn to master their own vocal instrument. Of course, physiology is to some extent a given: some people have larger resonators or longer vocal cords that produce a deeper voice, while sheer physical size, fitness, lung capacity and social confidence all come into play. But whatever you are given at birth and whatever road you have traveled to the here and now, you can become a master of your voice and make the most of the equipment you uniquely own.

WHAT YOU CAN DO

If your voice is important in your life (which I believe it is for everyone), find a coach and get to work on it. Search the internet for local experts—try terms such as "voice coach," "singing coach," "public speaking coach"—then call a few that you like the sound of, book trial sessions, choose the one you like best and work with them to improve every aspect of your voice: posture, breathing, projection, timbre, prosody, pace, pitch, body language and gestures—all the tools in your vocal toolbox that I described in my 2013 TED talk *How to Speak So that People Want to Listen*.

The Elizabethans described language as "decorated silence." How would the world be if we all had the intention of decorating the silence when we speak? Your voice is unique, and you probably haven't scratched the surface of what it can do. I urge you to develop this skill. You will not regret becoming a masterful speaker: it will transform your outcomes in life.

Storytelling

Though the tradition of engaging group imagination through storytelling has largely been displaced by more solitary and passive entertainments such as books, films and gaming, there are still professional storytellers in the world, and plenty of other people striving to keep the skill and the old stories alive. Edinburgh hosts the Scottish Storytelling Centre, which trains aspiring yarn-spinners and runs events, including an annual Storytelling Festival, which is the largest of its kind in the world.

Orkney's best-known storyteller is author Tom Muir. I asked Tom where storytelling fits today, in a world of handheld devices that can do almost anything. He replied:

> I think people want something a bit deeper, because a lot of modern culture can tend to be rather shallow. People who are maybe more thoughtful, more insightful, have a different way of looking at the world. And I think this is something that not only interests them, but I think there is a need for it. It's a two-way street because what you're trying to do as a storyteller is to paint pictures in people's minds—always avoiding the risk of overworking a piece. You don't want to stand there and give them minute descriptions of things, because that gets boring, and also people want to be able to form those pictures in their own minds. Your idea of somebody or someplace will be completely different from the person sitting a few inches next to you. There's also a desire among many people to have a folk identity. It's the stories that their ancestors have passed down, something

that is traditional to their culture, their people, their tribe—and I've used the term "tribe," meaning us, everybody, because we all have our own tribe.

Another accomplished storyteller is Mike Sowden, whose Substack newsletter *Everything Is Amazing* entertains thousands. He isolates a vital and increasingly rare function of storytelling: maintaining ambiguity in a world where everyone has an opinion. He told me, "A story can say some very complicated and ambiguous things, and it can leave them hanging in a way that's satisfying for the story, but it doesn't resolve in your mind: you go away and feel that this is a problem that nobody has solved, and now it's my turn to think it through. That's a real gift, a thing to treasure, because everything is uncertain. That's what science teaches. When you form an opinion, everything collapses into one form and it's incredibly difficult to break that up again and allow yourself to rethink something. Scientists have working hypotheses, so how about having a working opinion that's subject to change?"

As well as stories, words spoken, chanted or shouted in unison can join people together, for good or ill. The largely cheerful banter in instantly created chants and songs at football matches creates a sense of identity and belonging, matching in sound the visual identifiers of team shirts, scarves or hats. Prayers or mantras said in unison are mainstays of many religious gatherings. Slogans have long been chanted by marching demonstrators, baying mobs or, most chillingly, by huge crowds at rallies pledging their loyalty to despotic regimes. This brings us to the other omnipresent human sound: music.

Music

Next to speaking, music is the most primal human sound. Every human society ever discovered, no matter how remote or cut-off, has developed music, so it arises wherever humanity exists. To be human is to be musical, although the reverse is not true, as we've seen: humpback whales and birds have developed wonderful songs of their own.

Let's not get entangled in the complex question "what is music?"—other than to enjoy two very different definitions that might challenge you to create your own. Italian virtuoso, composer and writer Ferruccio Busoni once said: "Music is sonorous air." And, as we have already seen, Hermann Hesse coined the lovely construction that, "Music is time made aesthetically perceptible."

A definition I strongly disagree with is Steven Pinker's "auditory cheesecake," which he used in his 1997 book *How the Mind Works*, writing, "As far as biological cause and effect is concerned, music is useless," suggesting that it is entirely hedonic and serves no other function. If this were true, music would surely not be so universal, or so essential in social situations—an argument well put by Daniel Levitin in his book *This Is Your Brain on Music*. It seems more likely to me that music was the progenitor of language, originating with mothers humming to their babies and the mimicking of sounds and rhythms experienced in nature, and then becoming the primary form of social bonding and communication in what Steven Mithen in his book *The Singing Neanderthals* called a "proto-hum"—wordless vocalizations that used timbre, pitch, rhythm and prosody to communicate emotions, needs and commands.

When language evolved, music lost its communication function, though throughout human history it has kept or developed several other functions of almost equal importance. Communal spiritual practice has centered on song from the earliest shamanic times to modern day hymns, worship songs, chants, Sufi *qawwali* and the muezzin's call to prayer. Social cohesion has always been expressed and strengthened in rhythmic chant or singing: just a few examples include the historic Māori *haka*; the rain dances of shamanic tribes in North America, Africa and China; laments at funerals; celebration songs at weddings, births, harvests or victories in battle; music at rites of passage—from initiations and graduations to confirmations and bar mitzvahs; and even today's supportive or cheerfully provocative chants by football fans the world over. Music has long been a tool of courtship—there are more songs about love than about any other subject—but it can also be a weapon: after the rebel Scots were defeated at the Battle of Culloden in 1745, bagpipes were legally classified in Britain as a weapon of war and remained so until 1996. They were used in many conflicts to terrify the enemy: such was the ferocious reputation of Scottish warriors, as late as the First World War, some 2,500 British soldiers served as pipers and crossed No Man's Land armed only with their bagpipes.

Humans naturally turn to music to express themselves, to join together or reach out to the ineffable. As Steven Mithen writes, "Without music, the prehistoric past is just too quiet to be believed." *Homo sapiens* existed for up to 200,000 years before language developed, and our ancestors for millions of years before that. It is inconceivable that they lived in silence for all that time. Steven Mithen's theory that music came first resonated with me as I watched our baby Sapphire, who, as I started this book, had recently learned to walk and did not yet have words. The first thing she did on finding her balance was to jiggle up and down and twist from side to side when she heard music.

There are still many places in the world where folk music thrives. In Orkney, the old music is strong and vibrant. It seems that every child learns fiddle or guitar and there are many bands available to book for the frequent cèilidhs; each summer, the Orkney Folk Festival brings together some of the world's best performers, and the jams between players from all over the world show how wide and deep the river of Celtic music is.

Someone who has probably encountered more global ethnic music-making than anyone since the legendary Alan Lomax is the musician, author and filmmaker Jamie Catto. In 1999 he persuaded Island Records founder Chris Blackwell to fund a seven-month global treasure hunt for him and his collaborator Duncan Bridgeman, where the aim was to meet, record and film iconic musicians and thinkers on four continents. Using just an Apple G3 Mac as their studio, they layered one field recording on top of another, adding Indian flutes and Māori singers to African drummers and Western rock stars, trusting in fortune to make each step the right one—as when they saw legendary Indian singer Asha Bhosle having tea in the garden of the Rambagh Palace hotel in Jaipur, approached her to contribute, and recorded on the spot a transcendent improvised vocal, which R.E.M.'s Michael Stipe then added to. Bridgeman mixed the hundreds of recorded sessions on the remote Scottish island of Eilean Shona, and the result was the multimedia DVD *1 Giant Leap*, nominated for two Grammys and spawning the global hit single "My Culture." Not content with that, the pair then set out on a second odyssey, this time taking four years and visiting fifty locations to assemble a further film called *What About Me?*

After all that, Jamie Catto should know better than most why music is so fundamental to people all over the world, so I asked him exactly that. He replied:

> It's the feeling of being in the river of genius and jubilation, the place where everything's all right, the place where we are all good, the place which is beyond all our petty stuff. It's like going to a funeral, you know: you might have a gripe with someone that you've had a falling out with. You go to a funeral, you see them, you hug them, everything suddenly seems petty in that place. Funerals and births and music concerts all lift us out of the everyday petty, egoic, self-cherishing nightmare into a place where we are with each other... In the villages, you have the natural fragrances of the flowers, the natural beat of the pulse of nature, the natural everyone singing together and being unified. Yes, we're all in individual egos, individual bodies, but there is a place where we all connect, and where that is most alive you will find music, because it is the expression of the divine, therefore the expression of our interconnectedness, and we want more of it.

Sci-fi author Kurt Vonnegut agrees, saying in the *1 Giant Leap* film: "Music is to me proof of the existence of God."

In the development of human music, singing and chanting must have come first, probably centered on communal shamanistic practices and, where possible, taking place deep underground. Reverberation adds depth and presence to singing: a singer in a resonant space gains extra power and gravitas from the lingering echoes compared to someone singing in the open air or in a dry room, which is why reverb is routinely added to vocals in recordings. Early humans sheltering in caves would have noticed this effect. There is little left of their cultures now, but we do have some evocative remnants in their cave paintings, which the archaeoacoustician Iegor Reznikoff discovered are usually found in the most reverberant places, while other resonant spots that were too awkward to paint in were marked with red dots. If resonance was that important to those people, it seems highly likely that ritual chanting went hand in hand with the cave paintings, some of which were almost certainly made by Neanderthals rather than modern humans. We will never know how far back in time music goes, or if proto-humans had music before *homo sapiens* existed.

Paleolithic tubular bone quena-type flute found at the prehistoric site of Geissenklösterle in Swabia, Southern Germany.

To accompany the voice, surely the first instrument was percussion, starting with clapping, stomping and slapping the torso or legs. It's a short step to banging on objects with sticks or bones, then discovering that resonant, hollow objects sound better than solid ones. Frame drums exist in all cultures, and my bet is that they were the first man-made instruments. However, a strip of wood and an animal skin are not very durable, so it's not surprising that the oldest musical instruments yet found are three flutes made of bird bone and mammoth ivory found in the Geissenklösterle cave in Germany's Swabian Jura region that are around forty-two thousand years old. One is pictured above.

Intention is very important with all sound, and music has always been intended to affect people. It still does, in all four of the ways we explored at the start of this book—physiologically, emotionally, cognitively and behaviorally.

Some music is simple, and its intentions are crystal clear, but sometimes composers may encode profound ideas or complex emotions into a piece. I called Benjamin Zander, conductor and music director of the Boston Philharmonic Youth Orchestra, after experiencing his first, extraordinary TED talk about the power of music, when he had the entire TED audience standing up and singing the choral finale to Beethoven's Ninth Symphony at the tops of our voices [56]. Before each of his orchestral performances, Ben gives a talk where he explains the ideas and structures of the piece, so that people can more deeply experience it. In our call, he gave me an example of how it helps to understand the intention of the composer: "There are two themes in the Beethoven Ninth: the secular (the brotherhood of man), and the sacred (the existence of a loving Father beyond the stars)... The moment when they're joined together [in the double fugue] sends shivers down my spine. That's why Beethoven's metronome mark is so slow: he wants this, the central idea of his life, to be powerful and rooted in a slow, marked way, not rushed through. I say all this in my explanation, so that people can appreciate not only the pretty tune and the fact this sounds quite complicated and nice, but also the idea that they can actually change their lives from it." Ben's lovely TED talk gives a heartwarming example of how this can happen, as understanding a sad piano piece finally liberates a Northern Irish youngster's tears for his lost brother.

We can all appreciate art at different levels, and we don't always have the time or inclination to delve deep into the maker's intentions in an art gallery or when watching a movie. But I have found that conscious listening is almost always rewarding: it will usually reveal hidden dimensions of any sound, enriching one's life experience and strengthening one's consciousness as a result. With music, adding some understanding of the intention and the craft of the composer, as well as the skill of the performer, can transform a superficial appreciation to something far deeper, more inspiring, and more meaningful.

For much of human history, everyone did this, because music was a communal activity. At some point, we created a distinction called "musician" before the rest of us took our seats in the audience. Now, our relationship with music is often even less intimate than that. As AI is deployed to create music of increasing quality, we will learn much about whether great music must always contain human creativity. I have to admit to a hope that, however brilliant AI music becomes, it will always fall short of achieving the emotional and spiritual connection that human creativity can achieve.

Reproduced sound

In the modern world, we experience most of our music not live but recorded—a category of sound that simply didn't exist prior to Edison's epoch-making phonograph in 1887. We've come a long way since then, and the pace of change has inevitably accelerated: the gramophone arrived in 1895, the microphone in the 1920s, magnetic tape and stereo sound in the 1930s, vinyl records in the 1940s, cassettes in the 1960s, CDs in the 1980s, MP3s in the 1990s and iPods in 2001. Now, everything is changing with digital streaming of hi-res music and spatial sound, two seismic innovations that are remapping the way we listen to music.

R. Murray Schafer coined the term "schizophonia" to describe the experience of experiencing sound without seeing its source—for example, listening to an orchestra on a stereo in your living room. Mobile devices and headphones have made this experience both now portable and ubiquitous, enabled by the rise of digital sound recordings.

Digital sound actually started almost a century ago. In the 1930s, Bell Labs invented pulse-code modulation, the method for digitizing sound that's still used today. It involves sampling the sound at rapid intervals, so its fidelity depends on the number of samples per second and how much information is in each sample, which is known as "bit depth." Four bits can just about manage a supermarket checkout bleep; eight bits can manage old-fashioned arcade game sound effects; sixteen bits is enough for good quality music.*

The problem with aiming for high fidelity in digital sound is that it creates enormous file sizes. When storage space was very limited in computers (and even more so in portable devices), the only solution was to squash the files by removing as much data as possible without compromising the audio quality unacceptably: this is called file compression, and it relies on understanding the foibles of our hearing, which is a discipline known as psychoacoustics.

---

* Something called the Nyquist–Shannon sampling theorem dictates that the sample rate must be at least twice the maximum frequency of the signal, so for human ears that can hear up to 20 kHz the sample rate must be at least 40 kHz. Eventually, the standards of 44.1 kHz and a bit depth of 16 bits emerged in the "Red Book" standard for CDs in the 1980s. In professional audio recording, the norm is 48 kHz (or more recently 96 kHz) at 16 or 24 bit.

This science started at the end of the nineteenth century, when the American physicist Alfred M. Mayer published a paper showing that a tone can be rendered inaudible by a lower tone. Since then, the challenge has been to identify the elements of speech or music that can be discarded without the listener hearing much of a difference. For voices in telephony, that's anything below 400 Hz or above 3,500 Hz. Music is far more complex, but psychoacoustics has developed to the point where it can identify the musical data that can be stripped out without you noticing too much: for example, a simultaneous bass drum and bass guitar note share many frequencies and you don't need them twice. Nevertheless, it took German electrical engineer and mathematician Karlheinz Brandenburg years to refine the compression algorithm that was eventually adopted in 1992 by the Moving Pictures Expert Group (MPEG) as MPEG-1 Audio Layer III or MPEG-2 Audio Layer III, both more commonly known as MP3. He used Suzanne Vega's song "Tom's Diner" as his reference track, listening repeatedly as he tweaked the code to ensure that the quality of her voice was affected as little as possible while he eliminated most of the data in the audio file to shrink it to a small fraction of its original size. As a result, he dubbed Vega "the mother of MP3."

Small MP3 files were the enablers for Apple's launch of the iPod and iTunes in 2001: suddenly, you could have your music collection in your pocket, though you had to rip the files from vinyl, tape or CD first. As Internet speeds increased, streaming was the next natural development; Spotify launched in 2006, challenging the whole paradigm of buying and owning a physical product. Apple, Amazon and Google/YouTube followed, all using very compressed music.

MP3 offers various levels of quality that depend on the bitrate, measured in kilobits per second (kbps). In the early years, many people ripped their music collections at the lowest bitrate of 64 kbps to cram as much music as possible onto an early generation iPod. They probably regretted it later, because the drop in quality is very noticeable: a CD has a bitrate of 1,411 kbps (44,100 samples per second, times 16 bits, times two for stereo), so a 64 kbps MP3 is missing a whopping ninety-five percent of the data. But higher bitrate MP3 files at up to 320 kbps offer reasonable quality and still take up only a tenth as much space as the original. MP3 is "lossy"— it actually removes audio—but later "lossless" compression formats such as Free Lossless Audio Codec (FLAC) and Apple Lossless Audio Codec (ALAC) preserve all the audio data and yet still cut file sizes by at least half.

The rationale was clear at the start: an hour of music in MP3 at 128 kbps takes up just 57 MB of space, compared to 630 MB for CD quality or a massive 4.2 GB for 24-bit 192 kHz hi-res audio. Today, lossless compression can shrink CD file sizes down to around 25 MB per track and under 400 MB for an hour of streaming.

Ryan Maguire from the University of Virginia's Center for Computer Music brilliantly showed what goes missing with lossy compression in a project called *The Ghost in the MP3*. Using "Tom's Diner" because it was the guinea pig for the development of MP3, he used audio spectra to compare the compressed version with the uncompressed and then extract the difference, which he assembled into a track called *moDernisT* (an anagram if you hadn't spotted it) [57].

Stanford University's professor of music, Jonathan Berger, tests new students every year by playing them sounds in a range of compressions formats, from MP3 upward, and sadly he found that their preference for the harsher, sizzling sound of low bitrate MP3s rose each year, presumably because that's what they became used to on their devices and budget headphones. This is an alarming trend to me, because the more data that's removed by clever psychoacoustic compression, the more work your brain has to do in order to imagine it's still there. There is no research I'm aware of to prove this, but I am convinced that listening to highly compressed audio is fatiguing and increases stress. Hopefully, the new trend toward hi-res music will reverse the preference for poorer quality MP3s.

Another modern trend is connected to shifts in the economics of music. Stung by the massive revenue loss due to the shift from physical products to digital streaming, the music industry and its channel partners, the licensing authorities, are strongly pushing the use of background music (known as BGM in the trade), with research that unsurprisingly "proves" that we all love music everywhere. However, the independent research I've seen, some of which is helpfully collated in a downloadable factsheet by the UK's Pipedown ("the campaign for freedom from piped music") [58], shows that for many (especially older people) it occurs as just more noise, which in the long term has physical, social and spiritual consequences. Particularly when played in reverberant rooms like coffee shops, through substandard sound systems, music on top of noise is just more noise.

Given that most music is made to be listened to, it's a dense, attention-seeking sound that rarely helps people who are trying to work, think or have conversations. It can be very beneficial where people are bored by repetitive or dull work, and wonderfully effective when deployed appropriately—for example, club music in fashion retail; a string quartet at a classy cocktail party; or the brilliant use of Leftfield's "Phat Planet" in the iconic Guinness "Surfer" commercial of 1999 [59]. But music is still the most abused sound: those who manage brands and public spaces need to think carefully about designing appropriate, effective and healthy soundscapes rather than mindlessly playing generic music for the sake of it.

Now we arrive at the activity where design and sound are most powerfully interwoven: the creation of imagined worlds.

The sound shapers

From its birth with the first "talkie," 1927's *The Jazz Singer*, movie sound has evolved to the point where the film industry is one of the most sophisticated and innovative sound-makers in the world. It was only in 1977 that *Star Wars* transformed the hitherto flat, mono movie experience with the launch of Dolby Stereo, followed in 1983 by THX's digital sound, then the Dolby multichannel systems of 5.1, then 7.1 surround sound, and now Atmos, which can place sounds precisely anywhere in an auditorium.

Well-designed sound is transformative for video. If you ever watch a film without its score and sound effects, you will discover how devoid of emotional impact it is. Why is this so different from classic theater, where the acting is sufficient on its own? Rather like the sound of a car door, this is consensual myth-making: car-makers know how to pad the hollow metal of a car door to produce just the right satisfying meaty clunk, which they know we will interpret as a sign of a quality car. We like it, and they deliver it. With films, we expect to be wowed by crystal clear, larger-than-life sound, and would feel hugely short-changed if a film dropped the make-believe and gave us only authentic sound with no enhancement. Our expectations are very different in a theater.

Through the years, the people creating film sound have developed an ever-larger bag of tricks to engage and direct our interest and emotions. Diegetic sound is internal to the film's universe: mainly dialogue and sound effects; non-diegetic sound is external: for example, a music score (other than songs in musicals) or a narrator's voice.

Even in diegetic sound, almost nothing is as it seems. Location dialogue can be hard to capture without mics being seen on camera, and quality can be variable, especially outdoors in a wind, so the voices you hear have often been recorded after the filming in a process called Automatic Dialogue Replacement, or ADR, where the actors watch loops of each line and recreate their performance in a sound studio.

As for sound effects, you can almost guarantee that nothing you hear is the actual sound of the thing you're looking at, thanks to Foley, a highly skilled discipline named after Jack Foley, who innovated the addition of enhanced sound effects by recording a jangling bunch of car keys to imitate Roman soldiers marching with their equipment for Stanley Kubrik's *Spartacus* in 1960. Footsteps in snow, hoofbeats, thunder, punches, gunshots, ships crashing through waves, fires crackling and swords clashing in battle will probably all be added by a Foley artist in a studio—crunching corn starch imitates footsteps in snow; wrinkling cellophane mimics crackling fires. Today, there are libraries of digital Foley sound effects as well as sound designers who can create new ones on request. For *Jurassic Park*, sound designer Gary Rydstrom created the roar of the T. Rex by slowing down and mixing together recordings of multiple animals, including baby elephants, alligators—and not to forget Buster, his Jack Russell dog.

Sound can herald the next scene in a technique called overlapping, where you will often hear the dialogue or sound effects from a scene before the visual transition takes place. It can create a pervading theme throughout a movie, as with the ominous, almost omnipresent sounds of water (and the removal of almost all other sound effects) in Peter Weir's *The Last Wave*, presaging the tsunami that ends the movie.

To understand more about this, I spoke to David Sonnenschein, musician, filmmaker and author of the book *Sound Design: The Expressive Power of Music, Voice and Sound Effects in Cinema*. David explained how film sound designers work on four different levels of listening:

> Number one being the source: where's the sound coming from? A car passes, you place a glass on the table, or you have footsteps, and we make it all sound very real. The second listening mode is more about the shape or the physical quality of the sound, which we identify in sound engineering or sound editing by volume, pitch, attack, timbre, speed, rhythm—all these more measurable things, which are really the tools we work with. Then the third level is: what does the sound mean? Words have symbolic meanings, but we also have the intonation of the voice giving emotional context; something like a siren has the meaning of "pull your car over," or there's an emergency, or there's a bank robbery and the police are coming. And then the fourth one is referential: the meanings have something to do with our memory. That can be within the film itself, an internal reference, but then we have external references that have expanded beyond that, like Darth Vader's breathing. That sound has become iconic for us, but before that film came out it was just a scuba tank respirator. So we've given a lot of meanings to sounds both inside the film and outside.

These days, the skills learned in movie sound design are also applied to a wide range of video-based communication on TV, YouTube, social media and the web.

Like many film people, David Sonnenschein has also worked in game sound. This now huge industry had humble origins—I remember to this day the tension-building, gradually accelerating four-note motif of the *Space Invaders* machine that my friends and I avidly played in the Mitre pub in Cambridge in 1978—but gaming has now grown to overtake the movie business as the world leader in innovative sound design. A modern game may contain hundreds of hours of content, all with responsive sound effects delivered in high-definition surround sound. Most game music soundtracks are as professionally produced as movie ones, and some have broken out to become real-world hits: there have been many orchestral concerts featuring game music, especially from *Final Fantasy* [60]. The makers of game sound effects, background music and ambiences are now using generative technology, where sound is created algorithmically in real time to match the player's actions and state. Artificial intelligence will speed up this process, making game sound a powerful, organic, responsive element of the gaming experience. If VR ever really takes off, spatial sound will also be assimilated, and game sound will undoubtedly be the leading edge in sound innovation and design. Which brings us to the essential question: what next? It's time to do some crystal ball gazing.

6

THE SOUND OF TOMORROW

The future is already here—it's just unevenly distributed.
William Gibson

Innovation in sound reproduction has leapt forward exponentially over the last few years. Let me explore a few strands of the "future already here" that I have encountered, and extrapolate from them into an extraordinary sonic future that I believe is just around the corner.

Digital sound is thankfully emerging from its initial low-quality feeding frenzy, where the aim was to squeeze the maximum number of songs onto a small device, and is now enabling pristine quality and spatial sound. The leading accusation made by purists against digital sound is that it loses depth and feels flat compared to old-fashioned analogue recording. In analogue recording, soundwaves are turned into continuous electrical impulses; though there may be noise like hiss introduced along the way, there is no sampling, so there are no gaps and nothing is missing. But higher sampling rates are now closing those gaps to the point where even extremist traditional audiophiles will be hard pushed to tell the difference between digital and analogue.

The next big thing in sound reproduction goes even further. It's called spatial audio, and at the time of writing is already available on the Amazon, Tidal and Apple streaming services in two formats: Dolby Atmos and Sony 360 Reality Audio. There are other formats such as DTS:X and THX Spatial Audio, but Dolby and Sony look set to duke it out for dominance in the global market in the way Betamax and VHS did for video tape. Sony lost that battle even though its Betamax product was undeniably superior, but it has high hopes for 360 Reality, which is based on an MPEG standard for multichannel audio. It currently requires specific (mainly Sony) headphones or speakers to sound its best, though Sony is negotiating with Google and others to license its technology. The other contender, Dolby Atmos, does not require special headphones, and starts with a lead because many home entertainment systems have multi-speaker setups that can already play Atmos movie sound. Apple has partnered with Dolby and added its own enhancement to the technology in the form of dynamic head tracking, so that if you are watching a movie on your phone while wearing Apple earbuds and you turn your head, the sound source will still appear to be coming from the phone's direction. I suspect that Sony may again come second best in this standard war.

In headphones, music recorded specifically for spatial audio sounds exciting, though remastered older tracks are highly variable; this probably depends on the skill and commitment of the person doing the remixing. Nevertheless, Eddy Cue, Apple's senior vice president of services and the manager of Apple Music, is certainly convinced that spatial, not hi-res, is a breakthrough akin to HDTV and will be the future of music. He told *Billboard*: "I think this is going to take over everything. It's the way I want to listen to music when I'm in my car. It's going to be the way I listen to music immediately with my AirPods. It's going to be the way I listen to music in my house. In a way, it won't feel very good when I'm listening to something that's not Dolby Atmos because it's so good. It's like when I'm watching HD, it's hard to go back."

Is Eddy Cue right, or will spatial sound go the way of 3D video, which launched with loud fanfares but is now seen by most as a fad that's being kept alive almost single-handedly by *Avatar* director James Cameron? My bet is that spatial sound is here to stay, largely because of gaming. Whether or not virtual reality ever really takes off, gaming cries out for spatial sound, and it's already been deployed: Dolby Atmos powers Microsoft's 3D Spatial Sound platform for Windows and Xbox, and clever psychoacoustics in the headphone version allow gamers to experience sounds coming from above and below them on games such as *Call of Duty: Warzone*—while, of course, Sony's PlayStation uses the company's own proprietary technology. As Dolby and Sony battle it out for market domination, perhaps our entire conception of listening to music will change.

Meanwhile, even more exciting technologies may change our relationship with sound forever. Let's move closer to the edge to see the dawn of a new age of sound.

New front ears

Bone conduction is a technology that works by transmitting sound vibrations along your cheekbones instead of through the air, bypassing your outer and middle ear entirely. I have some bone conduction headphones today which are great for conference calls, because they manage voices very well and leave my ears completely free. I wouldn't use them for music, but the technology will doubtless improve, as it always does. The quality of bone conduction sound depends on the strength of the physical linkage, which is light with current wearables. Also, a lot of processing will need to be applied in order to experience sound in full bandwidth; with digital sound, that shouldn't be too difficult. I firmly believe that bone conduction will be the future of sound reproduction.

Someone who's already tested the limits of bone conduction is Neil Harbisson, the human cyborg and radical artist. I met and interviewed him at TEDGlobal back in 2006. Neil was born unable to see color, so in 2004 he had an antenna implanted in his head, sprouting from his occipital bone at the back, and curving over so that its sensor stands above his forehead. The device transposes the frequencies of light (which are all contained in a single octave) down forty octaves to become sound vibrations that he perceives through bone conduction as the antenna resonates his skull. This means that Neil *hears* color: he can discern thirty notes in each of the twelve semitones in the octave. The device is not limited to visible light: uniquely among humans, Neil can "see" ultraviolet and infrared too. He was surprised to discover that human faces are not black, brown or white, but all shades of orange, and he has scanned the faces of many famous people and created musical versions of them. It works the other way for him too: he perceives sounds as color, so he paints music.

Even more extraordinarily, Neil's antenna is connected to the internet, so he can receive color remotely. When we last spoke, he told me, "There's five people in the world that have a connection to my head, one in each continent. I suddenly feel that I'm perceiving something that someone else is perceiving. So, if someone is looking at the sunset in Australia, this person can use his mobile phone to send colors directly to my head. And then here in New York, I suddenly feel that I'm perceiving a sunset, and it feels like I'm sharing a sense with someone, and it's nice!" He is surely the first human being to perceive someone else's sensations remotely and directly.

Similar technology exists in surface transducers—devices that resonate any rigid surface like a sheet of glass or a wall, floor or roof, and turn it into a loudspeaker. These have been around for some time, resonating shop windows, bus shelters or museum display cases to deliver high-quality sound with no visible source. The principle is the same: sound transmission by conduction through a solid medium, which, as we know, is far more efficient than through air.

Meanwhile, in-ear technology is moving fast. I have worked several times with leaders in the hearing aid industry and seen how they are transforming the life experience of their millions of customers with modern electronics and digital sound processing. App-controlled modern hearing aids are smart and good enough to produce high-quality sound by recognizing and adapting to different situations—from conversation, music and nature sound to a crowded room. These companies are pushing the frontiers of in-ear sound technology, which (in the form of earphones or buds) now has the largest share of the leisure and music market, displacing traditional on-ear and over-ear models, especially among the young. Wearable audio is a huge and fast-growing market, which brings us to the next item on my future-facing list: voice-user interface.

Billions have been invested by Amazon, Apple and Google among others in speech recognition and voice synthesis, resulting in the explosion of voice-activated smart speakers answering (in the world outside of China) to Alexa, Google or Siri: over 200 million were snapped up worldwide in 2023 alone, and over half of all homes in the U.S. and UK have at least one installed. In our homes and in our cars, we are getting used to managing devices and making internet queries by voice, liberating our fingers from typing or touching screens, and our eyes from looking at text or graphics. Until now, voice bots have been relatively dumb, managing conversations by picking responses from pre-programmed databases, but they have yet to fuse with the fourth and possibly most powerful future strand: AI.

AI is already having a seismic effect on sound in communication. I've done some work with Davit Baghdasaryan, the CEO of Krisp, an extraordinary AI-based sound utility that intelligently removes background noise and unpleasant acoustics on calls, leaving only the speaker's voice and producing a clean signal that's almost studio quality regardless of what's actually going on around the caller. Similar AI-based technology is built into the Descript app, which is revolutionizing audio and video editing and allows its users to create an AI-generated version of their voice that can read printed material and even be configured to speak with different moods.

Weaving together these strands of the future, I have for years been forecasting an age of voice-user interface (VUI)—starting with in-ear devices incorporating AI and creating a virtual, in-ear assistant we can converse with in natural language, a little like J.A.R.V.I.S. from the early *Iron Man* films. This assistant will hold all our preferences, history, contacts, financial tools, diary and projects in its cloud-based memory and will act as our gatekeeper for people, businesses, organizations, government and any other external contacts we have any connection with. Apps on phones will be gone, other than the odd time we want to look at visual content when we're not in range of a screen we can throw it onto. We will interface with the world, via the internet, by speaking and listening, returning us to the days before text-based communication took hostage of our eyes and our fingers. That to me is a tremendous benefit, though the big unanswered question here is: who will own that intelligent agent and the data it can access? Will we have control of how it sounds and what it knows, or will that be one or more of the big tech companies? The latter raises all sorts of dystopian and even Orwellian specters.

As we edited the text for this book, my forecast became a reality for the first time in a TED talk by Jason Rugolo, founder and CEO of Iyo, which developed from Google's "Moonshot Factory" development initiative [61]. Rugolo demonstrated the Iyo One, an audio computer worn as two small discs with silicon-molded earpieces that not only listens and responds conversationally, but also allows its owner to alter what they hear by, for example, eliminating background noise in a busy room when trying to listen to a conversation, or even translating from other languages in real time. This may or may not be the device that moves us all into VUI, but it is at the very least the dawn of this new age, when our eyes and fingers will be liberated and our heads can come up from looking at screens.

I believe the next inevitable step from in-ear VUI will be skull implants using technology like Neil Harbisson's, so that we can speak and listen to our personal J.A.R.V.I.S (or personal music, or anyone else we choose to let in) using pure bone conduction. We won't need Neil's antenna, of course, so these devices will be invisible, aside from some sort of charging port, and our relationship with the internet will become a new sixth sense.

We will need to develop robust safeguards around this technology, though, especially in terms of age restriction. Descript is scrupulous about the generation of its artificial voices, but there are plenty of less principled developers creating deepfake apps that can use even brief recorded samples to create a believable facsimile of anyone's voice. As always, the denizens of the dark side invest more, and faster, than those trying to stem the corrupting flow, which now also includes highly realistic video deepfakes.

For recorded audio and video, a dystopian future is almost upon us: reputations are going to be vulnerable to forged "recordings" that provoke baying mobs on social media and are hard to disprove; at the same time, recorded audio or video that's not somehow watermarked or of unimpeachable provenance will soon be worthless as evidence, while those who really are caught on mic or camera doing something wrong will have bulletproof deniability in most cases.

We can only hope that some genius will invent an incorruptible watermark, perhaps using blockchain technology, so that the forgers and malefactors will be stopped in their tracks. This is urgent. Thanks to the internet, we already have conspiracy theories and fake news galore, and malevolent deepfakes could be seriously destabilizing to civil society.

Meanwhile, however, advances in audio technology are improving our life experience in shared spaces in some more prosaic but very significant ways.

Public sound

When I used to commute into London's Waterloo station in the 1980s and 1990s, the public address system was shocking: geriatric horn-type loudspeakers suspended from the ceiling of the cavernous, glass-roofed concourse would rasp out crucial information about delayed trains or platform changes, all of it almost completely unintelligible. The main issue was the vast distance from the speakers to our ears: the confusing echoes were as loud as the original signal. If you've ever flown from a large airport, you've probably experienced something similar. ("Was that us? What gate did they say?") This is all changing, thanks to small line arrays—long, thin units housing multiple loudspeakers that can be discreetly installed on pillars or walls at ear level, eliminating that pesky echo and delivering a much better-quality signal in the first place.

Large public spaces and concert halls are also benefiting from another innovation in pro audio: beam steering, which allows sound engineers to focus the sound where it's needed in order to reduce echoes and eliminate dead spots in any space, ensuring that everyone can hear every word or note.

I first experienced this when I visited Meyer Sound in Berkeley, California. John Meyer is an earnest, articulate sound pioneer whose long, straggly beard and ponytail bear witness to his lengthy stint learning his trade in the 1960s and 1970s by innovating sound systems for (and touring with) the likes of the Steve Miller Band, Creedence Clearwater Revival and the Grateful Dead. Founded in 1979 by John and his wife Helen, Meyer Sound now has over a hundred patents for audio innovations, and supplies sound technology to an illustrious client list, including the likes of Disney World and Ed Sheeran. I was stunned when John demonstrated the accuracy of his column array loudspeaker, which can project a beam as narrow as five degrees.

The Meyers are also innovators in spatial sound with their Constellation technology, which can transform the sound of a space at the touch of a button by, for example, switching a room between the acoustics of a nightclub, a cathedral and a grand concert hall. I asked John if this wasn't a little disorientating: surely this is a whole new level of schizophonia? He replied, "If you project a picture of a cathedral when you turn the room to 'cathedral,' it's more believable… it doesn't take much of a picture. In our theater, a ten- or fifteen-foot screen with a picture of a cathedral is enough: it just seems normal. In one of the installations we have in San Francisco, the younger conductors and composers are thinking about changing [the acoustics] as they play, which pushes our technology into a dimension that we hadn't thought about. So it's a creative tool as well as a listener experience." The auditorium may be about to join the studio as an instrument to be played, with dynamic, creative acoustics supporting a performance and amplifying its effect.

And perhaps not just auditoria, because the Meyers are not alone in pushing the boundaries of sound in spaces. L-Acoustics has long been an excellent manufacturer of sound systems for live events. In 2024 it launched its HYRISS (hyperreal immersive sound space) system that can change any room from a spa to a concert hall at the touch of a button, delivering distortion-free sound from multiple speakers embedded in every surface. I have experienced this system and it is astounding.

Beam steering technology is now also being applied to microphones. Shure bowled me over with a demo at their European headquarters in Eppingen, southern Germany, of their conference room steerable array microphone, a discreet ceiling-mounted panel for conference rooms that will automatically sense the location of the person speaking around a large boardroom table and pick up that voice in a narrow beam, muting everyone else. This is orders of magnitude better than the old-fashioned speakerphone in the center of a large table (usually in a room with terrible acoustics) that made conference calls tortuous and stressful experiences.

The tightest sound beams are found in what's known as directional sound. A little earlier we met HyperSonic Sound, which wowed me when Woody Norris demonstrated it at TED in 2004. Installed in museums and art galleries—or, like its competitor, F. Joseph Pompeii's Audio Spotlight, in theme parks—this technology can create "wow" moments with very focused pools or beams of sound, though it predictably lacks any bass: it's excellent for birdsong, running water or voices, but very poor at music.

And it's to music we now turn, as we consider the impact of technology on our listening to, and creation of, this important sound.

The death of music?

When I see people listening to music on headphones while reading a book, it makes me wonder, first, how do they do that? (I simply can't hear music and do much else at the same time.) And, second, what's the future of music if it is turning into a kind of veneer on life experience? Most music is made to be listened to, which is why it's not the best background sound in many situations. But perhaps our relationship with music is changing, much as our human relationships have with social media—becoming wider and shallower.

When I was a teenager who owned a few dozen LPs, each painstakingly chosen after a lengthy bus ride and an equally long deliberation in my nearest Our Price record store, every album was a treasured friend that I knew intimately. I grew up listening to vinyl records where the songs were in the order the artist wanted and you had to get up and turn the record over halfway through. Today, many people don't even have the patience to finish a song before flicking on to the next in a playlist. At school, I could look cool by reading *Melody Maker* and tucking the latest hot album under my arm. Today, my favorite music interface, Roon, offers me a selection of several new albums every day. I skip through tracks and if I like one, it might get a few listens before the next new batch deposes it. Even this curated selection is the tip of the iceberg: there is more music created in a day than I could listen to in a year.

I do worry that the supply glut and the ease of obtaining a song at the touch of a button are combining to erode the value of music. And has our attention span diminished to the point where an album is an archaic concept? I called renowned music producer, DJ and songwriter Mark Ronson to gain his perspective on my concern that we are devaluing this most precious of human sounds.

Mark's reply was encouraging: "I think the reason that the demand for vinyl is there is not just because of this kitschy nostalgia, but because people do want to sit down again with a record and listen. There's so much noise in our everyday lives, there's so many places where music is involuntary and you have no choice but to listen to music—it could be while you're on hold, it could be walking through a store. It could be just the fact of this knee-jerk pendulum swing against the scattered 'ADD playlist' generation. But it's kind of wonderful to watch, not just because I'm happy for the return of vinyl, but it gives me hope."

Mark's hope comes from seeing a move away from what he calls the dispensable digital space and toward more traditional, physical interactions—for example, the trend among Zoomers away from Kindles and back toward real books. He recalled a conversation with

David Byrne about the importance of variety in human experience: "[David] had the same thing, like cooking in the kitchen: you start chopping an onion and there's a record on, and there's something so lovely about having to go through the whole thing from start to finish and not necessarily loving every track; there's dips and valleys." The modern process of making an album is very different, and Mark feels it's at least partly to blame for falling sales: "You have a big pop star going in with ten of the hottest producers, all trying to outdo each other for who's making the loudest or the best song, who's going to have the biggest banger—so those records became emotionally and aurally exhausting. Putting on an old record, where it's meant to have ebbs and flows, and the shitty song makes the next song that you love even more enjoyable, I think that's such a lovely thing about the process of listening to a record, and being able to listen the way the artist you loved intended it."

I hope Mark's right and that this is a swing of the pendulum back to attentiveness—though I still can't help but worry about a world where so much of our modern experience of such a primal thing as music has become solitary. Making and listening to music were once intrinsically social experiences, but now so many of the thousands of tracks being created every day are made by individuals on computers to be consumed by individuals on headphones.

I think the key to the audio revolution shaking down and the cream rising to the top is intelligent search. We need trusted guides—curators we can rely on to do the research and present things we will find interesting, plus AI-based algorithms that will actually make engaging, appropriate playlists or selections that introduce appropriate, high-quality content to each listener: the current machine-based selections offered by Spotify, Pandora and the rest veer far off-piste in a fairly short time. This need for trusted guides is why radio is doing so well now and has a bright future, despite media fragmentation. Also, the globalization of the music market does mean that a radio station, album or podcast focused on a tiny subgenre that would have struggled to attract a hundred customers locally a few decades ago, now becomes a viable proposition with an audience of thousands.

Technology has always driven the forms of music. Once, it was wandering musicians on the Silk Road who brought esoteric sounds from other cultures; now, everything is available to anyone on the internet, and this universal melting pot is in danger of replacing folk idioms and traditions with standardized global subgenres. Also, virtuosity is no longer required to make successful music, though it remains a useful asset in some genres: a Paganini, Tatum, Rich or Satriani will always be in demand.

In music creation, digital signal processing allows creative people to manipulate sound like never before: whether that's making it easier to do traditional things like adding reverb, echo and other effects; creating entirely new techniques such as autotuning vocals, emulating exact spaces through convolution reverb, or slicing, stretching and pitch-shifting sound; or opening up whole new creative vistas with algorithmic sound and programming languages like Ableton's Max.

Many people are investigating computer-generated music, whether that's generative sound as pioneered by Brian Eno (with Tim and Pete Cole) in the 1990s, which uses probabilities to trigger samples and create endlessly shifting patterns in sound—or, more spookily, AI music, which learns tricks from recordings and churns out variations on those themes or replicates the voices of stars alive or dead. How good is it? You can make your own mind up by exploring the latest proponents on the internet; at the time of writing, Melodio [62] is active and seems popular, but the scene is fast-moving and will have changed by the time you read this. For my money, AI can't yet write a good hook but, of course, this is the dawn of the technology and we have yet to see how scary, powerful and perfect it can become at impersonating (or replacing) human beings in music. Another AI engine, Suno [63], seems able to create serviceable songs with vocals in almost any genre to a brief from the user, even including names of loved ones. Doubtless, soon AI will have moved on apace and will be able to make music that's hard to distinguish from human-made.

However, to me, good music has a *purpose*: it conveys meaning or emotion from the creator to the listener. It is unclear what purpose an AI neural net can have, apart from clever impersonation or pastiche, and I share Nick Cave's concern that using AI to replace human creativity is very dangerous—as he says, we are "fighting for the very soul of the world." Unfortunately, technology is unstoppable, and Google, Spotify and many start-ups like Suno are developing AI-based deepfake music tools which will inevitably result in huge issues in copyright, artist remuneration and reputation. For example, Steve Marriott's children were vocal in their opposition to his widow's plans to release "new" songs using AI to generate vocals.

A much more positive result of technology is that for creating and performing music, digital sound has made possible instant, perfect quality sampling of any sound—including bits of other people's work—to be looped, deployed or adapted into a new creation. Mark Ronson's TED talk on sampling [64] is a fascinating exploration of this phenomenon in music-making, which has always relied on people standing on the shoulders of their predecessors.

Another seismic change that is affecting the lives of professional musicians the world over is disintermediation. Once, the record companies reigned supreme but now, thanks to the internet, musicians are regaining control and ownership of their work and forging direct connections with their audiences that help to short-circuit the very distant (and low-income) relationship offered by the streaming services. Imogen Heap posts every step of the creative process for her followers and, like many artists now, has her own app and offers levels of monthly subscription that fans can choose between. There are many platforms offering to help artists promote, distribute and sell their music—or, even more relevant in an age where digital assets have little perceived value, their relationships, where increasing levels may yield limited edition exclusive books, videos, photos, unreleased audio and even private performances.

Perhaps online sharing services and the direct artist–fan relationships made possible by the internet will return music full circle to its primal functions of defining and cementing communities, and bringing people closer together.

Back to nature

A final promising trend in the sound we listen to is the recent recognition that many of the sights, sounds and smells of nature are good for well-being, reducing stress and even improving concentration in workplaces. In architecture and interior design, the increasingly popular trend of biophilia is reconnecting people with nature when they are indoors, and improving the hermetically sealed artificial environments in which we now spend over ninety percent of our lives. Some common expressions of biophilia include multiple plantings and even green walls, pictures or videos of natural scenes, soundscapes of nature, and ways of maximizing natural light and airflow. This approach is fashioning some cities too, through the work of architects like Stefano Boeri, with his vertical forest towers in Milan, Thomas Heatherwick with his *1,000 Trees* project in Shanghai, and the Vietnamese architect Vo Trong Nghia with his cascading gardens.

Biophilia can be very effective in sound: as noted, the research-based work of Evan Benway and his team at Moodsonic is transforming the workplace experiences of thousands of people around the world, replacing distracting noise with scientifically designed, pleasing natural sound that also does a good job of masking unwanted conversation and helping solo workers to concentrate without stress. Importantly, this sound is generative—created live in real time by computer, using probabilistic algorithms, so it doesn't loop and repeat patterns that people might recognize and find irritating.

Biophilia is particularly timely, because the pandemic caused a major rethink of the whole office paradigm. Many people discovered the benefits of working from home, while organizations discovered the huge savings in time, money and carbon of virtual meetings. In the future there will need to be good reasons for people to congregate in offices—mainly to meet, rather than to do solo working—and these spaces will have to be much higher quality, more akin to high-quality hotels or airport lounges, in order to attract people to travel in.

The Leesman Index is the world largest data source on office conditions, and Leesman's CEO, Tim Oldman, told me, "The very distinct data trend is that the higher the quality the experience of the workplace, the more employees want to spend time in it." Professor Jeremy Myerson, author of numerous books on workspace, agrees. When I asked him about the future of the office, he said, "We're not going to travel in every day. We're going to go in for big project kick-off meetings, for training, for mentoring, for social activities, for team-building activities. Offices are going to become a bit more like hotel lounges. They're going to pay much more attention to air quality, to acoustics, to standards of furnishings: they're going to go upmarket and the food's going to be better. It's going to be much more of a curated experience. I think this is where offices and the hospitality sector begin to merge to some extent." Sound will be an important element of this change, so perhaps noisy offices will at last be a thing of the past.

Many structures built in the last few millennia have made use of one particular natural sound: running water. Alongside their practical, aesthetic and cooling functions, fountains are often symbolic: water can mean life, cleanliness, purity, flow, nature—and, in private dwellings, wealth, because a fountain is beyond the means of most ordinary people, especially in hot countries. With the invention of aqueducts, and then siphoning to move water and manage pressure, the ancient Greeks and Romans created fountains in public places to celebrate leaders, victories or deities, and the tradition was revived, first by early Muslim architects and then in the Renaissance, as people rediscovered the tools and techniques of the ancient engineers. Many centuries-old fountains are still enjoyed today—for example, the three magnificent sculptures in Rome's Piazza Navona, or the beautiful fourteenth-century Fountain of Lions in the Alhambra Palace in Granada, Spain. When powered pumps became available, some of the world's most iconic fountains came into being, including the twin fountains in London's Trafalgar Square, and the world-famous Jet d'Eau in Lake Geneva. Nowadays, fountains can be much more dynamic: they dance, choreographed by show-control programs on computers and synchronized with (often deafening) music, as at the Dubai Fountain in the Burj Khalifa Lake. Costing $218 million, this circles back to one of the earliest reasons for having a fountain: to flaunt wealth.

The sound of running water can affect people in many ways: it can soothe, refresh, energize or enhance well-being, as well as having a practical function in masking unwanted sound, especially conversation, to create privacy. In Japan's long tradition of Zen gardens, water, or *mizu*, is an important feature, and the sound of it dripping into basins, flowing through streams or pouring from bamboo fountains can symbolize the flow of time, as well as purification and cleansing. Two water-based design elements of Zen gardens are forms of sound sculpture. The first is the *suikinkutsu*, an upturned pot with a hole in its bottom, often placed below ground and concealed by rocks or pebbles, onto which the overflow from a basin drips, creating an unpredictable tuned sound akin to wind chimes, but with no visible source so that it directs the visitor's attention to the intangible and invisible [65]. The second is the *sōzu* or *shishi-odoshi*: originally designed as a deer-scarer, this is a bamboo tube that gradually fills with water until its center of gravity shifts and it tips over on its hinge, emptying its water out and then clunking on another piece of bamboo as it swings back to its starting position—it makes a satisfying hollow, woody sound that echoes around the garden at regular intervals [66].

Hopefully, the new focus on biophilia, well-being and quality in architecture, along with the exciting developments in technology, will start to produce healthier, more productive sonic environments for us all.

WHAT WE CAN DO

Humanity is a noisy species. Once, our noise was a minor irritant but, as with so much of our impact on the planet, the issue is one of scale, now that we are eight billion strong and multiplying constantly. There are very few places in the world now not touched by our din, as we travel, build, exploit resources, socialize or fight. The constant noise in populous areas, especially cities, is having disastrous effects on the health and happiness of the people living there, even if many are unaware of it, and we are deafening many species on land and especially in the oceans, creating huge issues for them in navigation, communication, feeding and breeding.

Sound can join us together, but it can also engender terrible destruction. Its power is still largely ignored—something I hope this book will help to change. We need to acknowledge and take responsibility for the sound we make, because it has profound effects on our planet—and on us. Despite the mountain of evidence about the harmful effects of our noise on ourselves and the rest of the world, I fear that tranquility will become an increasingly rare and precious thing, eventually limited to only the most inaccessible places or the wealthiest people.

*

Now we move on to explore sounds that have been here far longer than life itself: the sounds of our planet Earth.

7

GEOPHONY

The three great elemental sounds in nature are the sound of rain, the sound of wind in a primeval wood, and the sound of outer ocean on a beach. I have heard them all, and of the three elemental voices, that of ocean is the most awesome, beautiful and varied.

Henry Beston, *The Outermost House*

On 27 August 1883, around midday local time, fishermen on the Andaman Islands, an archipelago in the Bay of Bengal, raised their heads at what they thought was the sound of distant artillery. Later, throughout the day, similar reports came from New Guinea, Western Australia and Rodrigues, an island near Mauritius. In all, there were fifty separate reports of guns firing from locations across the globe, spanning an area of around 15 million square miles. What they were hearing wasn't a distant war, but the loudest sound in recorded history: the eruption of the Krakatau volcano, now better known by the Victorian misspelling of Krakatoa.

Earthquakes had been happening in the area for years previously and had started more than three months beforehand with an ash cloud and ongoing activity creating high tides and masses of floating pumice. On the afternoon of Sunday 26 August, the volcano started its last phase of destruction. As continuous eruptions intensified, pumice and ash fell and the skies darkened for many kilometers around. Powerful static electricity discharges struck vessels, people and buildings; airborne shockwaves shattered windows and stopped clocks; and the sea became increasingly violent, with dangerous tsunamis lashing the coastal villages. Some people were lucky in the most bizarre ways: a powerful wave lifted one man in his bed and deposited both unharmed on a nearby hilltop. The constant roaring was audible as dull thuds in Singapore, over a thousand kilometers to the north—then, at 4 a.m., the pattern changed to discrete thumps "like a steam-engine, emitting full-throated *whoomphs* as it gathered speed."

The four final, gigantic explosions started at 5:30 a.m., culminating in the ultimate cataclysm. At 10:02 a.m. a huge tsunami left the island, then two minutes later Krakatoa obliterated itself in a titanic explosion estimated to have been the equivalent of 200 megatons of TNT—ten thousand times more powerful than the atomic bomb dropped on Hiroshima in 1945.

The sound power level of this detonation at source has been estimated to have been at least 310 dB, and the Batavia gasworks in North Jakarta, 160 kilometers away, registered a sound pressure level spike of over 8.5 kPa, equivalent to 172 dB; the sound was so loud that it ruptured the eardrums of half the crew on the Royal Mail ship *Norham Castle*, over sixty kilometers away. It is extraordinary to conceive of a sound so loud that you could not make yourself heard when the source is hundreds of kilometers over the horizon. The eruption's shockwave circled the globe seven times at over 1,000 kph.

Pyroclastic flows including toxic gasses carried by a cushion of super-heated steam raced outwards for up to forty kilometers at speeds of 100 kph. These and the airborne shockwave killed anyone in their path, but the bulk of the 36,147 victims of the eruption died from drowning, as two huge tsunamis destroyed countless towns and villages on the coasts of Sumatra and Java. By lunchtime on the Monday, the island of Krakatoa had completely disappeared, though its legacy continued to affect the entire world for months. The temperature in the northern hemisphere dropped by half a degree the following summer, acid rain fell worldwide, and spectacular sunsets inspired artists for months.

The victims of Krakatoa had little or no warning of what was about to happen. Today, we have equipment that can listen for the early sounds presaging huge eruptions like this, giving people more time to escape from the deadly aftermath. This is one way in which more attentive listening can save many lives.

*

We humans can be a noisy bunch, but the sounds created by our planet tend to operate on a whole other scale. Our planet creates an infinite range of sonic textures, from the zephyr of softly moving air or the trickle of gentle summer rain to the most violent events. It hums all the time and is never quiet. There are sounds we can savor and appreciate to enhance our quality of life, and others that are dire warnings or terrifying consequences of cataclysms that have claimed millions of lives. It behooves us to listen carefully, both in order to connect with and better understand our home planet, and also to be more sensitive to warnings of disasters to come. To understand how and why, we need to go back to the birth of our planet.

Roughly 4.6 billion years ago, part of a vast dust and gas cloud, not unlike today's Orion nebula, condensed. Its center collapsed as gravity drew mainly hydrogen atoms together into a massive sphere; eventually the pressure became great enough to start nuclear fusion, and our Sun was born, lighting up a vast disc-shaped cloud around it. The Sun absorbed 99.9 percent of the material in the cloud but couldn't take the last 0.01 percent: the disk was spinning, and its angular velocity created enough centrifugal force to counterbalance the Sun's gravity. Gradually, through uncountable collisions and the inexorable work of gravity, that spinning 0.01 percent coalesced into clumps of matter and then small "planetesimals" up to ten kilometers in radius—some of which then grew to be much larger, forming up to a hundred planetary embryos in the inner solar system alone. For millions of years these danced around the Sun, jostling one another, colliding and even melting from inner radioactivity, until they became the planets we know today.

The Earth is thought to be a composite of more than ten embryos. Its birth, like that of a human baby, was perilous, dramatic—and noisy. The process involved many violent events, including (according to current thinking) a truly titanic impact about 4.5 billion years ago, when the proto-Earth collided with Theia, a Mars-sized planet. This catastrophic encounter caused the two bodies' cores to merge, while ejecting enough matter to go into orbit around the merged planet and eventually form our Moon.

Most planetary embryos probably had thick atmospheres of helium and hydrogen, while the early Earth, its surface entirely covered in a magma ocean 2,000 kilometers deep, was shrouded in carbon dioxide, nitrogen and water. Sound had a medium to travel through, so an imaginary, indestructible microphone would have picked up deafening and thunderous sounds from these mind-boggling events.

Over the next few hundred million years, the Earth cooled, but explosive bombardment by asteroids and comets continued, possibly delivering much of our planet's water. The magma became solid rock and oceans formed, initially containing water far hotter than 100°C, kept liquid by the great surface pressure at that time. The Earth's surface conditions during this Hadean age were not unlike those on Venus today, with temperatures of 230°C and an atmosphere comprising mostly carbon dioxide. Nevertheless, familiar sound was surely already happening, with wind, thunder, rain, waves and running water. And somewhere on an ocean floor, probably in an anaerobic hydrothermal vent where magma and water met, life began.

The ground beneath our feet

While we think of the land as motionless in comparison to the sea, it is in fact extremely dynamic if we extend the timescale from human to geological. The layer of cooled crust on which we live is only about one percent of the Earth's volume. Like a thin eggshell, the crust is wrapped around the mantle, a layer making up eighty-four percent of the planet's volume that behaves like a viscous liquid, sometimes described as the consistency of caramel, which itself contains a core with temperatures at the center that match those of the surface of the Sun at 5,500°C. And our thin layer of eggshell is cracked, comprising seven or eight major tectonic plates (debate continues about which number is correct) and multiple minor ones. These plates effectively float on the mantle. The tremendous heat and pressure cause convection currents like those you can see in boiling liquids, and these currents cause the plates to move at a rate of about five centimeters a year. Sometimes they move toward each other, crumpling the thin layer of crust upward into mountain ranges. Sometimes the plates are moving away from each other and the resultant change in overlying pressure creates a pathway for molten rock known as magma to rise up and erupt through the surface of the Earth, where it becomes known as lava. Volcanoes often form along these seams, known as plate boundaries.

Krakatoa is the most famous volcanic eruption because it happened in an age of instant communication: the invention of the telegraph some fifty years earlier made this eruption an instant global front-page sensation. The 1815 eruption of Mount Tambora on the island of Sumbawa in Indonesia had been even larger, with a Volcanic Explosivity Index (VEI) of seven: Krakatoa was a six on this logarithmic scale, so Tambora was judged to be ten times as powerful. Tambora certainly killed more people than Krakatoa: scientists estimate that 12,000 people died in the eruption itself, and a further 49,000 soon afterward from famine and epidemic disease in the region, such was the destruction and toxicity Tambora spewed forth, killing crops and poisoning water supplies—its ash cloud created pitch blackness for two full days everywhere within a radius of six hundred kilometers. Tambora's massive ejections into the atmosphere caused a devastating cold spell across the entire northern hemisphere: 1816 was named "the year with no summer," and for three years crops repeatedly failed, livestock died, famine was widespread and there were many outbreaks of epidemic typhus. The widespread misery is said to have inspired Lord Byron's poem "Darkness," as well as Mary Shelley's *Frankenstein*.

Tambora's global death toll is unknown, but was probably several hundred thousand, making it among the worst natural disasters in history. However, we have no reliable measurement of the noise level of the eruption, which lasted several days. As with Krakatoa, it was heard far away, though the greatest recorded distance was 2,600 kilometers, to Sumatra, where people reported hearing gunfire, compared to the extraordinary 4,776 kilometer distance that Krakatoa's sound carried.

In 2021 the Hunga Tonga–Hunga Ha'apai eruption was almost as violent. This massive event caused an incredible 400,000 lightning flashes on 15 January alone, which must have created an overwhelming constant barrage of thunder. The eruption has since been confirmed as the biggest explosion ever recorded in the atmosphere by modern instrumentation. A sensor in Idaho, 10,000 kilometers away, registered six hours of infrasound (too low for human hearing) at 100 dB; a day later, the sound recurred there, having circled the globe. The Tongan eruption was certainly the loudest since Krakatoa, and we can hear it, thanks to a video posted on Twitter by Portia Domonatani Dugu, who was filming with her phone in Lakeba, Fiji—over 800 kilometers from Tonga—when the sound of the explosion arrived, almost an hour after it happened [67].

Widening our timescale, geological evidence shows that the Earth has suffered many eruptions much larger than any of these from supervolcanoes—and even bigger Large Igneous Provinces. Krakatoa ejected about twenty-one cubic kilometers (21 km_3) of rock into the atmosphere, and Tambora possibly as much as 150 km_3. The Youngest Toba supervolcano eruption in Indonesia around seventy-four thousand years ago had a VEI of eight (the maximum on the scale) and ejected 2,800 km_3 of rock in a single inconceivably enormous event that lasted two weeks. This devastating eruption is thought to have created a bottleneck in human population: it is thought that its catastrophic impact on climate, vegetation and water left less than ten thousand humans alive. Genetic evidence seems to corroborate this thesis, indicating that all today's humans descend from between a thousand and ten thousand breeding pairs that existed around seventy thousand years ago. If this is true, most of humanity died in fourteen days with the thunder of that terrifying event in their ears.

LIPs are longer-lasting events that can output over one million km_3 in an active period of volcanism lasting a million years, dramatically altering the Earth's geography and global conditions: the Siberian Traps LIP (uncannily centered on Tunguska, where the famous meteor later exploded) is thought to have warmed the Earth by a full 7°C.

Along with their other drastic global consequences, all these events would have created vast amounts of thunderous sound. With titanic supervolcano eruptions like Toba or Yellowstone, this would probably have rivaled the noise of the largest meteorite impact events like Chicxulub. Although the vast majority of these events predate humanity, one can only imagine the terror suffered by our ancestors seventy-four thousand years ago when Toba erupted, without doubt making the loudest sound ever experienced by human beings. What these distant ancestors thought they were hearing is beyond our imagination.

While LIPs happen only around once every twenty million years (the last one was seventeen million years ago), supervolcanoes are much more frequent. Yellowstone erupted two million years ago, then 1.2 million years ago, then 600,000 years ago—so it's about due to blow again. Humanity almost died out in the last supervolcano, so we can only hope that we don't hear the sound of Yellowstone, or any other supervolcano, for many thousands of years to come, by which time (if we have survived) we should be a spacefaring species able to escape the destruction. Meanwhile, the more and better we listen with seismographic and acoustic sensors, the better we will understand our planet's bouts of violence and be able to prepare for them.

There are around 1,500 active volcanoes and somewhere between fifty and seventy eruptions a year worldwide, but this number of geological noise-making events is dwarfed by another that occurs as tectonic plates grind sideways against each other, the friction creating enormous energy which is then released as seismic waves that cause the ground surface to shake. While earthquakes are nothing like as loud as volcanoes, they are far more frequent and create plenty of audible sound that's quite terrifying for those caught up in them, plus massive amounts of low-frequency sound that can be detected thousands of kilometers away.

Around half a million earthquakes take place every year—more than a thousand a day—but eight out of ten are too weak to be felt. Like the VEI scale for eruptions, the famous Richter scale (now supplanted by a more accurate scale known as moment magnitude or Richter magnitude) is logarithmic, with each increase of one indicating a tenfold increase in the amplitude of the seismic waves (unlike the VEI, this scale goes to ten). The strongest earthquake ever measured on a seismograph was the 1960 Valdivia earthquake in Chile, which lasted for ten minutes, reaching around 9.5 on the moment magnitude scale and generating tsunamis that were still ten meters high when they reached Hawaii, over ten thousand kilometers away. This earthquake killed around six thousand people, but in recorded history there have been many far more devastating earthquakes: for example, the 1556 Shaanxi earthquake in China may have killed over 800,000 people. With human population growing all the time, and massive cities like Tokyo, Jakarta, Istanbul, Los Angeles, San Francisco and Tehran sitting on or near major fault lines, the human cost of earthquakes is bound to grow exponentially in the future.

There are no professional recordings of earthquake sound or measurements of the sound pressure level during one, for obvious reasons: the events are short-lived, and people have far more important things to focus on when they are happening. The Seismological Society of America has made Karl V. Steinbrugge's compilation of mostly poor-quality recordings freely available on its website. The most powerful is the number 16 from Friuli, Italy, where an LP playing Pink Floyd skips because of the initial vibrations, and a second or two later the roar of the main earthquake is very audible [68].

Alongside the main terrifying thunderous roar, the crashes of objects and structures smashing and the screams of terrified people, earthquakes generate infrasound at 1 to 5 Hz and seismic waves at much lower frequencies. When the Honshu earthquake (magnitude 9.0) struck Japan on 11 March 2011, scientists of the Vents Program at Pacific Marine Environmental Laboratory using a hydrophone of the U.S. Navy's Sound Surveillance System recorded infrasonic soundwaves that had traveled over 1,400 kilometers across the Pacific. The sound was below the human audible range, but when pitch-shifted up four octaves, it becomes visceral and intimidating; toward the end of the recording, the massive acoustic energy overloads the hydrophone, completely distorting the audio [69].

Just as cosmologists use pitch-shifting to make low-frequency sound audible, a seismologist has developed a system to make seismic waves audible, helping us to perceive and understand the way earthquakes work. Zhigang Peng is Professor of Geophysics at Georgia Institute of Technology's School of Earth and Atmospheric Sciences, and he has created sounds (and sonograms) derived from seismic data by speeding it up hundreds of times. One example is a sound file created from the 2011 9.0-magnitude earthquake in Tohoku-Oki, Japan [70].

In another form of seismic sonification, seismicsoundlab.org took data about the size and depth of earthquakes and applied sounds and colors to them to create fascinating audiovisual representations of decades of earthquake activity in the world, and in chosen locations such as California or Sumatra. There are some stunning videos to enjoy [71].

There are good reasons for listening to the vibrations and sounds of earthquakes: maybe there are infrasonic indicators that will eventually allow prediction, or at least some advance warning. The numerous reports of strange animal behavior before earthquakes might be due to magnetoreceptive animals sensing electromagnetic waves from deep in the Earth—but to date there is no repeatable method of predicting where and when a quake will strike.

Not all sounds produced by the earth are cataclysmic, of course. Loosened by erosion, rocks and stones slide and fall from mountains or cliffs, and there are currently at least forty places in the world where sand dunes whistle or boom as sand shears down the slip face. This tuned sound can either be a quick whistle at 500 to 2,500 Hz, or a loud booming or roaring sound at 50 to 264 Hz that can last for several minutes. People, including Marco Polo in the Gobi Desert in 1295, have noticed and recorded this phenomenon for hundreds of years. There are booming dunes in Asia, Africa, the Middle East and the USA (mainly Nevada and California). You can hear a recording of singing sand from the Great Sand Dunes National Park in Colorado [72].

In 1923 Marquess Curzon of Kedleston published *Tales of Travel* and described the sound thus: "First there is a faintly murmurous or wailing or moaning sound, compared sometimes to the strain of an Aeolian harp...Then as the vibration increases and the sound swells, we have the comparison sometimes to an organ, sometimes to the deep clangor of a bell...Finally, we have the rumble of distant thunder when the soil is in violent oscillation."

There is another planetary source of extremely powerful noise—this one arising not from beneath our feet, but from above our heads.

Our planet's greatest hits

According to NASA, around 44 tons of meteoritic material falls on the Earth every day, almost all of it vaporised in the atmosphere as shooting stars, silently lighting up the sky every night. But some much larger chunks of space rock don't burn up safely: they hit the surface at thousands of kilometers an hour, creating the loudest explosions the Earth has ever experienced, and devastating consequences for living things. In the last 500 million years alone, we know of five massive extinction events, and at least twenty smaller ones. The most common causes have been meteor impacts, rather than supervolcanoes.

The last major extinction event was sixty-six million years ago, when an asteroid around ten kilometers in diameter, traveling at twenty kilometers per second, struck what is now Mexico's Yucatán Peninsula in a titanic impact that gouged a hole over twelve kilometers deep, now known as the Chicxulub crater, the second largest on Earth with a diameter of around 150 kilometers. The explosion, estimated at over seventy teratons of TNT, was fourteen hundred times more powerful than the largest nuclear device ever detonated, the USSR's Tsar Bomba. Winds of over 1,000 kph powered out from the blast, along with multiple megatsunamis up to 1,500 meters high, moving at hundreds of kilometers an hour, plus a shockwave that would have been lethal for over a thousand kilometers in all directions and a blast radius a thousand times more powerful than Hiroshima.

As the shockwave degraded into sound, it would have ruptured the eardrums of any animals within many thousands of kilometers, and possibly even on the far side of the Earth, and it would have circled the planet multiple times.

The longer-term effects of this explosion were even more devastating, as the impact ejected as much as twenty-five trillion tons of material into the atmosphere. Some of this escaped into space; some fell back to Earth in flames, setting fire to three-quarters of the forests on the planet, while a dense fog of the finer dust stayed in the atmosphere, cutting off the Sun for many years and drastically changing the climate, as well as creating rain that acidified the oceans. The result was the extinction of around eighty percent of all living creatures, including all the non-avian dinosaurs.

Such huge impacts are very rare, but the Earth still suffers strikes today: as recently as 2007, a three-meter-wide meteorite smashed into the ground near the Peruvian village of Carancas, near Lake Titicaca, shattering windows a kilometer away. And a hundred years earlier a meteor created the second-loudest sound in recorded history, though this one left no crater.

On the morning of 30 June 1908 an asteroid about fifty meters in diameter entered the Earth's atmosphere, traveling at over 50,000 kph. It exploded before hitting the ground, fortunately over a sparsely populated region of Siberian Russia near the Tunguska river, which leant its name to the incident. The detonation, the equivalent of somewhere between ten and thirty megatons of TNT, flattened eighty million trees over an area of 2,150 square kilometers; people twenty kilometers away were knocked off their feet, and windows were broken hundreds of kilometers distant. At source, the sound of this explosion was probably around 300 dB; the noise carried for great distances and the strange sounds of the event were particularly striking in eyewitness accounts, as reported in the *Krasnoyaretz* newspaper of 13 July 1908:

> Kezhemskoye village. At 7:43 the noise akin to a strong wind was heard. Immediately afterward a horrific thump sounded, followed by an earthquake that literally shook the buildings as if they were hit by a large log or a heavy rock. The first thump was followed by a second, and then a third. Then the interval between the first and the third thumps was accompanied by an unusual underground rattle, similar to a railway upon which dozens of trains are traveling at the same time. Afterward, for 5 to 6 minutes an exact likeness of artillery fire was heard: 50 to 60 salvoes in short, equal intervals, which got progressively weaker. After 1.5–2 minutes after one of the "barrages" six more thumps were heard, like cannon firing, but individual, loud and accompanied by tremors. The sky, at the first sight, appeared to be

clear. There was no wind and no clouds. Upon closer inspection to the north, i.e. where most of the thumps were heard, a kind of an ashen cloud was seen near the horizon, which kept getting smaller and more transparent and possibly by around 2–3 p.m. completely disappeared. *N. V. Vasiliev, A. F. Kovalevsky, S. A. Razin, L. E. Epiktetova (1981). Eyewitness accounts of Tunguska (Crash). Archived 30 September 2007 at the Wayback Machine, Section 1, Item 5*

Is a major, life-threatening meteor impact likely to happen again? The past says yes: the Earth Impact Database, maintained by the University of New Brunswick's Planetary and Space Science Centre, currently lists 190 impact structures on our planet. Though the largest ones were formed reassuringly far in the past, the most recent are just a few years old. We do now have powerful computers making it possible for NASA's Planetary Defense Coordination Office to track all potentially hazardous objects, so we would have plenty of notice of an impending Chicxulub-sized impact. And if we saw one coming, the 2021 Double Asteroid Redirection Test mission, which deflected the 150-meter-wide asteroid Dimorphos by crashing a probe into it, showed that we might now have the technology to divert a life-threatening object that's on a collision course for Earth, given sufficient warning.

Also from the sky above come some much more common planetary noises—far less dangerous and actually rated by many people as a favorite sound.

Thunder

As warm air rises, it cools, forming small droplets of water. Sometimes those droplets continue to rise and create larger droplets, which turn into ice crystals, rising through the cloud, accumulating more water droplets until they are too heavy to be carried by the updrafts and they fall as hail. As a hailstone falls, it picks up a negative charge, rubbing against smaller positively charged ice particles, which rise to the top as they are smaller. This is exactly the same sort of charge that builds when you rub a balloon on your hair.

Eventually, the attraction between the positive and negative becomes too strong and there is a flash of electrical discharge. Sometimes this is within the cloud and sometimes cloud-to-ground. A typical lightning strike travels at 350,000 kph and releases a gigajoule of energy, enough to power a typical house for almost a month, instantly heating the surrounding air to 30,000°C, which is around five times the temperature of the Sun's surface. This sudden heating causes the air to expand violently as a shockwave, creating the sound we call thunder. At source, the sound pressure level is typically 160 to 180 dB, but some lightning strikes can create a crack that's louder than 200 dB. People unlucky enough to be close to a bolt of lightning can suffer ruptured eardrums, as well as internal injuries from the shockwave and vision damage if they are looking directly at it—or, of course, death, if the bolt actually hits them.

The sound of thunder is categorized in four types: a clap is very loud and under two seconds long, with higher-pitched sound involved; a peal changes in pitch and loudness; a roll is irregular; and a rumble can last for up to half a minute.

Many people love the sound of distant thunder (ideally listened to from under a warm blanket in a comfortable, watertight house), so it's no surprise that there are myriad YouTube videos offering just such sound, also available in higher quality on the MyNoise website [73]. However, I doubt the people living by Lake Maracaibo in Venezuela are such big fans: that place suffers huge thunderstorms up to 160 nights a year, with each storm averaging twenty-eight lightning strikes per minute and lasting up to ten hours—which would mean forty thousand thunderclaps in one night! One recorded storm in Merrillville, Indiana, on the night of 17 July 2003, exceeded even that rate: apparently it yielded 104,280 lightning flashes, with over ten thousand hitting the ground. That's 158 lightning strikes and thunderclaps per minute for ten hours. I doubt many people slept in Merrillville that night.

Globally, there are three million lightning strikes every day. If you want to see where they are happening, lightningmaps.org offers a real-time map of the world, showing all the current activity [74].

Lightning is the most common cause of another much more destructive sound that has become all too familiar over the last few years: the roaring and howling of an uncontrolled bush or forest fire. Although some are started by human malevolence or carelessness, fires have been happening on the planet for countless millennia. Lightning, volcanoes and meteor impacts have all in the past created huge fires. Over the last few years, with extreme climate swings creating severe droughts, we have seen terrible fires raging on the U.S. West Coast and in central Europe and Turkey, among other places, with devastating loss of property and life.

A huge forest fire has a terrifying voice, as you can experience in a disquieting video. In 2019 the U.S. Forest Service planned a controlled fire at Fishlake National Forest, Utah, and researchers at the Pacific Northwest Research Station and Rocky Mountain Research Station saw the opportunity to gather vital data. They set up monitoring equipment, including a video camera in a special fireproof housing. You can watch and listen at [75]: not for the faint hearted.

Stormy weather

I doubt I'm alone in saying that one of my favorite sounds is a gentle breeze gusting through leaves or long, dry grass. It feels refreshing and reassuring, especially on a hot day: perhaps we innately associate it with new life and energy, since air is one of the three essentials for life [76].

Weather gives rise to many of the most familiar sounds that our planet has been generating for millions of years, since long before there were any ears to hear them. The Sun heats areas of the planet unequally, especially water and land because land changes temperature much faster than water—though, of course, different types of land and areas of the oceans also heat at different rates. Hot air rises, pulling cooler air in to fill the space vacated; this process occurs constantly at every scale from a few meters to thousands of kilometers across the planet, creating everything from welcome zephyrs in a desert to the fastest wind ever recorded—a gust of 407 kph on 10 April 1996 during a tropical cyclone at Barrow Island, off the coast of Western Australia. Extreme tornadoes generate even faster wind speed, exceeding 500 kph, but they rarely pass over an anemometer, and if they did, they would almost certainly break it. On 3 May 1999 a mobile Doppler radar unit estimated the wind speed above ground of a tornado near Bridge Creek, Oklahoma, to be 511 kph.

Outside of powerful storms like this, the windiest spot in the world is the peak of Mount Washington in New Hampshire's White Mountains, which held the world record at 372 kph (recorded in 1934) until that Australian cyclone gust pipped it. With global warming heating the entire atmosphere, and thus increasing the energy in the world's weather system, any such records are likely to be short-lived in coming years. Overall, however, the Antarctic is the windiest region on the planet, with sizable areas enduring katabatic winds that sweep down slopes at around 300 kph.

You do have to love wind to live in Orkney. We are very familiar here with the upper regions of the Beaufort scale, to the point where Orcadians dismiss the frequent force-eight gales as "just a puff." There are few cozier feelings than being tucked up indoors with the wind howling and moaning around the edges of the house outside. But how does the wind make those sounds?

Wind itself makes no sound at all until the moving air meets an obstacle—whether that's the pinnae on either side of your head, a tree, a power wire or a building—at which point it creates a series of turbulences. For a cylindrical object like a wire, these turbulences are regular and their sequence is known as a "von Kármán vortex street"; it has a specific frequency, and if that matches the resonant frequency of the wire, it will start to vibrate—hence, singing wires in wind. But for an irregular object the vortices are shed more chaotically, and the vibrations that reach the ear are often untuned, arriving as a broadband whooshing sound.

Historically, Japanese *kenjutsu* masters have focused in their swordplay training on achieving perfect *tachikaze* ("sword wind")—the sound of the sword through the air. A perfect strike involved the angle of the blade being identical to the angle of the blow, which would create tuned sound, or Aeolian tones; a pure whistling sound meant that the technique was perfect and the blow would cut through the target, while a swishing sound indicated that the blade was tilted compared to the strike angle, so the blow would be more glancing and less effective.

Aeolian tones are also responsible for the moaning sound as a wind wraps around the corner of a building or passes over a hollow object—they are well known in ships' rigging, where the varying speed of the wind can generate different harmonics, making the ropes sing a melody of sorts. This same effect is harnessed intentionally in Aeolian harps, where tuned strings await a breeze to create pleasing sounds, all of which are harmonics arising from the particular eddies in each string's von Kármán vortex street—resulting in a performance which is perhaps the ultimate in aleatoric music [77]. The young science of aeroacoustics studies all forms of sound created by air flowing over surfaces, with much of its focus understandably on ways to reduce aircraft noise by improving the acoustic design of both airframes and engines.

Water music

Since the Earth's crust solidified, it has been resonating with a whole range of vibrations that scientists call ambient seismic noise, or free oscillations, popularly known as the Earth's hum. Most of this noise is very low-frequency seismic waves, but some of it just intrudes into the bottom end of the human audible range and some people have reported hearing it. Scientists have known of this noise for well over a hundred years, but its source has only recently been largely identified: most of it is created by the complex interactions of ocean waves.

As the wind blows across the sea, it creates what are known as ocean gravity waves—the ones we're very familiar with seeing on the sea, moving more slowly than the wind, with a frequency of much less than 1 Hz: typical waves might break on a shore every six seconds, which is a frequency of 167 millihertz (mHz). More powerful infragravity waves can occur as subharmonics of the gravity waves; they have much longer periods of at least thirty seconds and can move far faster than the wind. As these surface sea waves collide, especially in winter storms, pressure waves then travel downwards at 1,500 meters per second; these impacts on the seabed, together with resonances in shallow continental shelves, set up constant seismic noise. This background sound includes microseisms (faint tremors) at around 140 mHz (mainly created by gravity waves), plus another seismic hum at the far lower frequency of under 10 mHz (a period of around a hundred seconds) that's generated mainly by infragravity waves and is as powerful as the maximum excitation from a magnitude six earthquake. The combination of these waves, traveling through the ground at over three kilometers per second, creates the Earth's hum, which constantly vibrates the rock under our feet.

Water covers around seventy-one percent of the Earth's surface. Though it accounts for less than a quarter of one percent of the Earth's mass, in total there is around 1.87 sextillion liters of it, including oceans, ice, lakes, rivers, groundwater and atmospheric vapor. (A sextillion is one with twenty-one zeroes, so a thousand billion billion.) As the endless global water cycle, powered by the

Sun, moves this vast reservoir endlessly through its three states, we encounter another primal noise of planet Earth: precipitation.

As with wind, there is enormous variety in the sounds of precipitation. My favorite sound of all is gentle summer rain on leaves outside a window [78]—a lovely sound that's emulated by many fountains—but rain (and its sound) can become far less pleasing when it's heavy or the temperature is low, especially when it's carried by strong wind.

If you like rain, the place to go is the Hawaiian island of Kauai, where it rains for up to 350 days a year. Some other famously rainy places are Seattle, Washington, and Bergen, Norway, where I remember a taxi driver complaining that the previous sunny day had stopped them from breaking their record for the number of consecutive days with rain, which was well over two hundred.

Hail is the noisiest form of precipitation, even in its mildest gravel-sized form. Hailstones form as strong updrafts of up to 180 kph in cumulonimbus clouds drive water droplets up to freezing high-altitude air. They are structured like onions, with multiple layers of ice forming around a cloud condensation nucleus, which is often a speck of dust, salt from the sea, or soot from human machinery. When they become too heavy to be held aloft by air currents, they fall, often over ten thousand meters, to the ground.

Most of the world's hail is harmless, making a characteristic rattling sound as it strikes roofs and windows, but larger hailstones can damage crops, vehicles, buildings and aircraft, and can be lethal to humans or animals caught without shelter—as happened in the ninth century in Roopkund, Uttarakhand, India, to hundreds of nomads who, archeologists have established, were killed by hailstones the size of cricket balls falling at over 160 kph. Even those lethal lumps of ice are small compared to the record hailstones recorded by the World Meteorological Organization: the heaviest was just over a kilogram, and the largest an incredible twenty centimeters in diameter. A hailstorm involving objects of that kind would be a truly terrifying experience, with sound akin to a pitched gun battle.

The subtlest and, to many, the most beautiful sound of precipitation comes when snow throws a blanket of silence over the landscape. In the right conditions, if there's no other noise and no wind, as the flakes fall and the tiny crystals melt in the perfect quietness, it's sometimes possible to make out the faint hissing sound, barely perceptible but extraordinarily enchanting. And, of course, a good fall of snow gives us the chance to experience a much-loved bit of anthropophony too: the crunching sound of feet treading in snow [79].

Moving water on the ground is our last stop on this tour of geophony (the sounds of our planet). Can there be many sounds more pleasing than the chatter of a babbling brook [80], the tinkling of a small waterfall [81], or the soothing cadence of gentle surf rolling on a sandy or pebbled beach [82]? In a 2012 survey in the UK newspaper the *Daily Telegraph*, the top three most-loved sounds were water-based: the sound of waves against rocks was top, with rain against windows second, and treading on snow third. The full list is in the Appendix at the back of the book, along with a poll of the most disliked sounds (the sound of retching tops that one).

Moving water has one of the widest sonic palettes of all, with many of the sounds having strong human associations. A dripping tap can be maddening; dripping water in an echoey cave speaks of darkness and mystery; while the sound of gentle raindrops can be contrastingly soothing and pleasing. In my talks, I often use the sound of water boiling in a kettle to exemplify everyday sounds that yield delight when we start to savor them instead of ignoring them: this sound has a story arc, dynamism, richness of timbre [83]—and strong associations with a nice cup of tea, a powerful positive overtone for us British!

Take a moment as you read the following lists of water sounds: stop and consider each one to see what associations you have with it. Some will be universal, and some personal, based on your own memories. At the gentle end of the scale are the sounds of water bathing, creeping, dribbling, dripping, drizzling, exuding, gurgling, hissing, lapping, leaking, meandering, melting, moistening, oozing, percolating, permeating, puddling, rippling, seeping, slopping, spilling, spitting, spreading, sprinkling, steeping, stirring, suffusing, tinkling, trickling, wandering, or weeping.

With moderate energy water may be babbling, breaking, burbling, cascading, coursing, dousing, driving, ebbing, eddying, flowing, flushing, irrigating, pouring, raining, running, sizzling, slapping, slopping, sloshing, sparkling, spinning, spitting, splashing, sploshing, spraying, springing, squirting, steaming, streaming, swirling, or washing.

Powerful water sounds include boiling, bursting, churning, crashing, deluging, drenching, drowning, drumming, engulfing, erupting, exploding, flooding, gushing, heaving, leaping, plunging, pounding, roaring, roiling, rolling, rushing, scalding, seething, shooting, soaking, spouting, spurting, surging, swamping, sweeping, swelling, tumbling, whipping, and whirling.

One sound of water moving from one state to another is particularly entrancing: that of ice melting in the polar seas. This is a highly symbolic sound now, as the Arctic ice fades away altogether and huge chunks of the Antarctic icepack calve into the southern seas, desalinating and raising the level of our oceans at a rate never before known in human history [84].

The loudest sounds of moving water come from two sources: waterfalls and waves. There are some massive waterfalls on Earth, and they can be deafeningly loud. However, the highest, Angel Falls in Venezuela, would not figure in any list of the loudest, because most of its water dissipates into mist during its 807-meter dive from the top of Auyán-tepui to the jungle floor below. And nor, perhaps, would the world's largest waterfall—because it's already underwater. In the Denmark Strait, between Greenland and Iceland, warmer surface water from the Irminger Sea sits on top of denser, colder water in the depths; this flows over the edge of a 3,500-meter undersea mountain and tumbles down to the sea floor below in a vast waterfall called in Danish *Grønlandspumpen* (the Greenland Pump). The flow rate is guesstimated at about five million cubic meters per second, which is an astounding two thousand times greater than Niagara Falls. The noise level is unknown, and trying to measure it would be very hard at that depth with that speed of flow—although water carries sound better and further than air, the noise will depend on the friction of the water over the mountainside and the steepness of the water as it reaches the bottom. With an angle far less than vertical, I think it's safe to assume that the noise level down there is not in the same league as the world's largest surface waterfalls.

The world's three largest waterfalls by volume are probably not the loudest, since they are configured in a long series of shallow-falling rapids on the Congo River. Taking the published figures for the top eight waterfalls for which measurements exist, I have multiplied the flow rate by the drop, then divided that by the width to get a rough proxy for the intensity of the noise.

	Flow (m/s)	Drop (m)	Width (m)	(f x d)/w
Guairá, Paraguay/Brazil	13,000	114	60	24,700
Paulo Afonso, Brazil	2,832	59	18	9,283
Kaieteur, Guyana	663	226	113	1,326
Inga, Democratic Republic of the Congo	25,768	21	914	592
Shivanasamudra, India	934	98	305	300
Livingstone, Democratic Republic of the Congo	25,060	6	701	214
Niagara, Canada/USA	2,407	51	1,204	102
Boyoma, Democratic Republic of the Congo	16,990	5	1,372	62
Para, Venezuela	3,540	64	5,608	40
Khone Phapheng, Laos	11,610	21	10,783	23

The clear winner is the Guairá Falls on the Paraná River on the Paraguay–Brazil border—sadly no longer in existence, because these mighty falls were submerged in the lake that formed when the huge Itaipu Dam was built in 1982. Prior to that, the roar of these wonderful falls could be heard thirty kilometers away. (A similar fate befell the Paulo Afonso Falls in Bahia, greatly diminished since damming in 1912.)

For anyone wishing to experience the stupefying noise of a vast waterfall that does still exist, Niagara is probably the most reachable. In former times, Niagara's thunder could be heard over forty kilometers away, but since the 1950s, half of the Niagara River's water has been diverted to power-generating plants; in winter months, another twenty-five percent goes the same way, so a summer visit is essential for maximum effect [85].

As with meteors and volcanoes, pre-human Earth experienced some even louder water movements, coming from mega-waterfalls, or immense torrents, when stone barriers holding back entire oceans gave way. In the USA Dry Falls in Washington State's Cascade mountains was once a waterfall twice as high and three times as wide as Niagara, making it the largest confirmed waterfall that has ever existed. But even this pales beside the event that refilled the Mediterranean six million years ago, known as the Zanclean megaflood. Tectonic plate movements had closed off the Med, which had partially evaporated as a result. The plates shifted again, and suddenly the door to the Atlantic was reopened. Scientists estimate that as the water thundered in, it formed a waterfall over 1,500 meters high that flowed at 160 kilometers an hour, raising the level of the Mediterranean by ten meters a day and refilling it in just a few years. Even more titanic events have surely happened several times in the Earth's 4.6 billion year history, though we have no geological evidence for them now.

This brings us to the last stop in our journey through the sounds of our planet: water waves. I live a short walk from Scapa Flow. It's one of the world's greatest natural anchorages, used for centuries by the Vikings, and then in both World Wars, forming the base for the British navy; it also has the entire First World War German fleet on its seabed after all those ships were scuttled in June 1919. This is well-protected water, and yet on the rocky beach where we walk our dogs we often note that great boulders have been shifted overnight by the power of the sea. On the exposed west-facing coasts of Orkney, the full power of the Atlantic pounds majestic cliffs, carving out sea stacks like the famous Old Man of Hoy, or gouging out "gloups," the local name for a collapsed sea cave or geo.

The sounds of sea waves are vastly varied: from ripples on a tropical coral beach within a lagoon, to the fresh sound of pebbles on a beach surging back and forth in a moderate swell, right up to the booming of huge ocean breakers smashing onto granite rocks—and the range of wave sounds doesn't stop there.

In certain conditions, tidal bores (large, fast-moving single waves) race up rivers like the Qiantang in China (up to nine meters high, traveling at 40 kph), the Amazon in Brazil (up to four meters high, traveling at 21 kph), Turnagain Arm of Cook Inlet, Alaska (up to two meters high, traveling at 20 kph) or the Severn in the UK (up to two meters high). In all cases, bores are loud, roaring as they roll, the air bubbles caught in large eddies generating a low-frequency rumble that's often heard far away. The engineer Hubert Chanson recorded a bore in the Baie du Mont-Saint-Michel in 2009 [86].

In conjunction with earthquakes and volcanoes, we've already mentioned the formation of terrifying tsunamis as enormous volumes of water are displaced. These walls of water can be tens of meters high when they come ashore. They may form trains of multiple waves with wavelengths of hundreds of kilometers that arrive up to an hour apart; they move at phenomenal speeds of up to 800 kph (reducing to around 80 kph as they reach a shore), and can cross entire oceans. They are also highly lethal: the 2004 Indian Ocean tsunami killed 230,000 people. Although most tsunamis happen in the Pacific—because of the combination of the "ring of fire" tectonic plate movements and the enormous volume of water—they have been known in the Caribbean and even in the Mediterranean: the 1755 Lisbon earthquake generated three tsunamis up to thirty meters high that struck the west coast of Portugal, as well as southern Spain and northern Morocco, killing sixty thousand people.

Along with the visual terror and the physical destruction, survivors report that tsunamis have a characteristic, equally awful sound. First, there is a sucking noise as the exaggerated drawback pulls the sea far from its normal position, after which many reports claim that wildlife goes silent. This is the few minutes' warning people have in which to get to high ground. Then, as the tsunami comes into view in the distance, witnesses report a crescendoing roar like that of a fast-moving train or aircraft. The wave rises in height as it reaches shallower water, due to friction at its base. It does not usually break: it just rushes onwards, carrying people, boats (and even ships), cars, buildings and trees far inland and pulverizing everything in its path, leaving behind a terrible soup of death and destruction. It then withdraws, sucking more people out to sea to their deaths, and there is a pause until the next wave in the train, which often fools people who had thought the destruction to be over.

Sometimes there are even larger waves, known as megatsunamis. These are mainly caused by landslips displacing water, just in the same way a rock thrown into a pond creates waves, but on a vast scale. The largest wave ever recorded happened in 1958 in Lituya Bay, Alaska, when an earthquake caused 80 million tons of rock and ice to slide into the water. The resulting wave was an incredible 520 meters high—seventy meters taller than Kuala Lumpur's famous Petronas Twin Towers, the tallest building in the world until 2004. The sound of such events is unrecorded and unknown: one can only try to imagine the roar of something a hundred times greater than the crashing of five-meter ocean breakers.

Humanity had a major hand in another recent megatsunami in Italy's Dolomite mountains, when a dam was unsafely constructed on the Vajont river, high above the Piave valley, despite expert advice. Numerous warnings from earth tremors and landslips into the reservoir above the dam were ignored, and early in the morning of 9 October 1963 a massive landslide created a 250-meter-tall megatsunami. Fifty million cubic meters of water overtopped the dam and fell hundreds of meters onto the sleeping villages below, obliterating them and killing around two thousand people. The sleeping people in those villages would have had just a few seconds of a booming or roaring sound as a warning before the torrent drowned or crushed them in their beds.

In pre-human times, scientists have established that there have been many far larger megatsunamis. The geological evidence shows that the Chicxulub meteorite impact sixty-six million years ago generated a megatsunami that towered a terrifying 1,500 meters high, followed by numerous waves at least a hundred meters high. Sediment records show that these tsunamis reached halfway around the Earth, and scientists estimate they had at least thirty thousand times the energy of the 2004 Indian Ocean waves. The sound of such an event is beyond our imagining.

WHAT WE CAN DO

We can all hope that we will never experience the lethal sounds of tsunamis, hurricanes or tornadoes, though the increasing heat in the planet's atmosphere and seas will inevitably mean that these events will be more common in the future, driven by greater energy in the weather system. It's important that we invest in research into understanding the sound of these destructive events so that we can better detect early warnings and hopefully thus reduce the deadly consequences, even if we can't avoid the events.

WHAT YOU CAN DO

There is great pleasure to be had from stopping to appreciate the gentler sounds of wind and water—and research is now proving conclusively that doing this is good for our health, reducing blood pressure and stress levels, even if it's recorded sound rather than the real thing. Experiment with the flavors of geophony that make you happy, or reduces stress, using sites like MyNoise or recordings on streaming services, and continue your journey to surrounding yourself with positive, healthy and productive sound.

8

COSMOPHONY

Sit, Jessica. Look how the floor of heaven
Is thick inlaid with patines of bright gold:
There's not the smallest orb which thou behold'st
But in his motion like an angel sings,
Still quiring to the young-eyed cherubins;
Such harmony is in immortal souls;
But whilst this muddy vesture of decay
Doth grossly close it in, we cannot hear it.

William Shakespeare
The Merchant of Venice, Act V Scene 1

In a clear, windless dawn on 22 July 2006, at NASA's Columbia Scientific Balloon Facility in Palestine, Texas, a vast helium balloon slowly inflated. Underneath the 821,000-cubic-meter canopy was suspended a 2.4-meter-tall vacuum flask containing 1,800 liters of liquid helium at just 2.7°K (kelvin)—less than three degrees above absolute zero, or minus 270°C. The helium was there to cool seven radiometers so that they could detect tiny radio emissions at centimeter wavelengths with frequencies in the 3 GHz region.

The whole extraordinary assembly was called the Absolute Radiometer for Cosmology, Astrophysics, and Diffuse Emission (ARCADE), a program run by a team led by Alan Kogut of NASA's Goddard Space Flight Center that was designed to search the sky for heat radiation from the first generation of stars, its wavelength having lengthened into radio waves as it traveled vast distances to Earth.

On this flight, the balloon rose thirty-seven kilometers to where the air is very thin and the instruments are high enough to get an interference-free view of the sky. For four hours, ARCADE's sensitive instruments recorded, then the cable to the balloon was cut, and the apparatus parachuted back to the ground.

When the team analyzed the data, they found something very strange: a "space roar" that brings into question our whole understanding of how the universe started. To understand why, we have to go back to the birth of the universe.

In the beginning

Sometime between thirteen and fifteen billion years ago, something happened, creating the universe we now inhabit. What this event was remains the subject of ardent debate and theorization by astronomers, physicists and cosmologists. We call it the Big Bang, but that's a misleading name because there was no audible bang at all.

Did the universe arise from a singularity—a tiny speck of unimaginable energy that suddenly inflated in a Big Bang? Or was it a big bounce from a pre-existing contracting universe? Both theories have supporters, though the Big Bang is far more widely accepted.

Scientists don't agree on when it happened either. The Hubble constant, the number that represents the speed of expansion of the universe and allows us to wind the clock back and estimate how long ago it all started, remains hotly disputed almost a hundred years after its definition. The oft-quoted 13.8-billion-year age of the universe is not yet an established fact.

Part of the problem is that the best tools of relativity and quantum physics all break down as scientists look back in time to the first instant of creation, less than 10^{-36} seconds (one trillionth of a trillionth of a trillionth of a second) from the start. As any physicist will admit, while brilliant minds have managed to deduce many of the basic rules underlying the development, structure and behavior of spacetime, matter and energy, there is so much we don't yet understand—including the dark matter and dark energy that appear to make up around ninety-five percent of the known universe.

What is generally agreed is that after a short period of extraordinary inflation the universe settled down as an expanding cloud of plasma. We know nothing about the shape of the cloud, or its extent, since we have no idea how large the universe is now, or what form it is. The universe we can *observe* is a sphere centered on Earth and roughly 13.8 billion light years in radius, but what's beyond our horizon is completely unknown. We certainly can't assume the entire universe is a sphere, or even that it has an edge at all.

After 100,000 years, each cubic centimeter of the universe contained about 2,000 protons and electrons and some 3,000 billion photons, all at a temperature of around 6,000°K (kelvin). It was a hot, thin, opaque plasma, appearing as a dazzling fog because, just as water droplets scatter light to make a terrestrial fog opaque, so in the early universe huge numbers of free electrons were scattering light and making the plasma opaque as photons bounced around, able to travel only tiny distances before bumping into yet another free electron. The electrons were free because at such high temperatures photons have very high frequencies, far into the ultraviolet—and photons with that much energy will tear apart any hydrogen atom they meet, separating its proton and electron. Hydrogen atoms were trying to form all the time, but their every effort was destroyed by yet another collision with a high-energy photon.

Yet in this primordial soup at the dawn of time and space, long before stable atoms and molecules could form, there was sound resonating through the fog.

Right from the start, tiny quantum fluctuations had generated a slightly uneven distribution of dark matter, and as a result areas of slightly greater density started to attract the surrounding plasma through gravity. As particles fell from all directions toward a denser region, the photons would compress and then bounce back, before falling in again and repeating the process. This continuous bouncing created spherical soundwaves, with frequencies varying from a few years to thousands of years, depending on the size of the clump of dark matter creating them. They were far too low in frequency for human ears to hear, but they were nevertheless genuine soundwaves.

Driven at almost sixty percent of the speed of light by the pressure of the densely packed photons, these sounds were loud, at over 100 dB, some of them with huge wavelengths of up to 700,000 light years and pitched some fifty octaves lower than the threshold of human hearing. The soundwaves generated a pattern of varied pressures in the plasma, reflecting inhomogeneities that are the basis of our universe today. Had the plasma been entirely homogeneous, there would be no matter, no stars and galaxies—and no us. As is found in many religious creation myths, the universe was effectively sung into existence.

Roughly 380,000 years after the Big Bang, the temperature of the plasma cooled to 3,000°K, and the universe started to become transparent as more and more photons cooled and shifted from their original ultraviolet, atom-smashing light toward cooler red frequencies. As this happened, more and more hydrogen atoms could hold together, and photons started to travel longer and longer distances. The fog didn't clear overnight: the transition to full transparency took around eighty thousand years. The photons that were released in that "decoupling" are still visible to our instruments today, at microwave frequencies, forming a sphere 13.8 billion years in radius around us that defines the edge of our observable universe. We call this radiation the Cosmic Microwave Background (CMB).

Here's a sonic metaphor that has helped my non-scientific mind to understand the CMB: if you were standing in a field in the middle of a huge crowd of people all shouting (that's the opaque plasma) and they all fell silent at the same instant, you would hear the people next to you stop shouting at once, but because sound travels at roughly 330 meters per second, you would hear people a kilometer away shouting for three more seconds before they fell silent; those ten kilometers away would take thirty seconds to go quiet, and so on. So, you would experience the last moment of shouting as a wave moving away from you at the speed of sound in all directions and getting quieter as it moved.

In the same way, the Cosmic Microwave Background is a snapshot of the "surface of last scattering" that's moving away from us at the speed of light (plus the speed of the expansion of the universe, but let's not get entangled in that), so the wavelength of the light reaching us gets stretched: its frequency falls from visible light to microwaves, a process that scientists call "red-shifting" because red light occupies lower frequencies in the electromagnetic spectrum. This is why the James Webb Space Telescope receives only infrared light as it searches for the most distant, ancient light sources in the visible universe.

In the extraordinary image of the Cosmic Microwave Background taken by the Planck telescope in 2013 [87], you can see tiny variations in the CMB temperature that are an echo of the varying densities of the plasma at the instant of decoupling. These tiny variations are what allow physicists to analyze the sound that was happening at that moment, and their pattern is now frozen for all time.

This is where ARCADE and its giant helium balloon come in. Although its data exactly matched previous spacecraft-generated estimates of the temperature of the CMB at 2.725°K, the measurements also revealed a glaring anomaly. The level of the diffuse radio background* was a full six times greater than that predicted by the prevailing theory of the creation, expansion and structure of the universe, which scientists know as the Lambda-Cold Dark Matter model.

"It's a diffuse signal coming from all directions, so it is not caused by any one single object," commented Kogut. "It also has a frequency spectrum, or color, that is similar to radio emission from our own Milky Way galaxy." This extraordinarily loud signal has become popularly known as the "space roar."

---

* The faint, omnipresent radio emission that can be detected throughout the universe. It is a form of electromagnetic radiation that appears to come from all directions in space, rather than from specific celestial objects. This background radiation is typically observed at radio frequencies and is thought to be composed of various sources, including distant galaxies, remnants of the early universe, and potentially other unidentified cosmic phenomena.

The source of the space roar remains a mystery. Scientists debate whether it's coming from within our galaxy or from somewhere (or something) else: speculations include turbulently merging clusters of galaxies, annihilating dark matter and supernovas of first-generation stars, among others. Wherever it's coming from, the space roar creates some serious issues. At a practical level, the noise obscures the signals from the earliest stars, making ARCADE's original mission much more difficult to achieve—and at a theoretical level, it is so out of line with the Lambda-CDM model's predictions that either the model is wrong, or there is something very significant going on out there that we just don't know about.

We've encountered the idea of pitch-shifting being applied to ultralow-frequency sound so it moves up into the range of human hearing, but it can also translate emanations like radio, microwave, light or x-rays, and even gravitational waves in spacetime itself, downwards into audible sound. This is creating *derived* sound to better experience and understand other forces.

There are at least two good reasons for doing this: first, it produces fascinating sounds that get us one step closer to things we can't otherwise directly perceive; and second, it often allows scientists to spot phenomena they would otherwise have missed. Sound is so visceral and quickly decoded by our brains that it's a great way to represent complicated data. In the same way that an experienced mechanic can listen to an engine and diagnose a fault deep within the machinery, scientists can now listen to radiation and vibrations from objects in the universe and more accurately analyze what's causing them. By using the technique of pitch-shifting, we can auralise the snapshot of the CMB taken by the Planck telescope in 2013, shifting the frequency up by fifty octaves to create something that we can hear. The audible version of the sound of the universe some 400,000 years after the Big Bang has a fundamental tone at 220 Hz (an octave below concert A, which is a tone of 440 Hz). This note is derived from soundwaves around 700,000 light years long that are compressing for the first time, plus a set of harmonics generated by other soundwaves that happened to be at maximum compression or rarefaction at the time of decoupling. Despite having harmonics, the sound is far from musical because the harmonics are broad: it sounds rather like brown noise (as discussed in Chapter Two) [88,89].

Mark Whittle is a professor in the astronomy department at the University of Virginia, specializing in active galaxies and black holes. He is also fascinated by the sound of the universe. He went one step further than just pitch-shifting the sound from the instant of decoupling: using computer modeling, he created a sound file that compresses the whole first 400,000 years of the universe into just ten seconds, again pitch-shifted up by fifty octaves so that we can hear it. This fascinating sound is a descending tone, because as the universe gets larger, the already enormous wavelengths can extend still further to hundreds of thousands of light years, so the frequency of the sound drops [90]. This sound of the primal universe is not just an interesting observation: it is woven into everything we now observe in the cosmos.

Following the release of the photons, there came a long blackness known as the cosmic dark ages. For over 100 million years, there was no light as the web of dark matter slowly and silently attracted diffuse hydrogen gas into pockets which gradually condensed... until, in the next landmark moment for the young universe, the pressure at the center of one collapsing gas cloud became great enough to start nuclear fusion and, somewhere in the darkness, the first star ignited. The stellar age was born, and once again light and sound existed.

As the millennia passed, countless stars ignited in groups that formed the galaxies and clusters of galaxies we now see. That structure is based on the framework of the web of dark matter, itself driven by the original inhomogeneities of the plasma fog that created those primordial soundwaves and their harmonics.

Speaking to me from his department in Charlottesville, Virginia, Mark Whittle described this wonderful link between sound and the cosmos:

> You might wonder whether the same harmonic patterns are present in today's universe, fourteen billion years later, in the distribution of galaxies. Do we see wave patterns in the patterns of galaxies? And the answer is: we do. It's difficult to see because the waves are so big: you need to survey a region of space about two billion light years in extent and measure the positions of about a million galaxies. But when you do that, and you analyze it...bingo! You find exactly the same harmonic patterns in today's universe. So those soundwaves got frozen back then, and they condensed into the galaxies and the same patterns are still present around us. And that was actually a beautiful scientific discovery because it really solidified this link between the present-day universe, filled with galaxies but otherwise dark and empty, back to the time when the universe was radically different—a brilliant, glowing, almost uniform gas with these soundwaves pulsing through it. And here you have a bridge: the soundwaves that you see on the microwave background are also witnessable today in the pattern of the galaxies.

Space is noisy

Most people think that space is a vacuum, but this is not true: it's actually a very dilute plasma—the same plasma that made up the primordial universe, now spread far more thinly. Aristotle was absolutely correct when he said, "Nature abhors a vacuum." At sea level on Earth, a cubic centimeter of air contains roughly twenty-five quintillion molecules—that's twenty-five with eighteen zeros—and each of those molecules contains multiple atoms. By comparison, in the cavernous voids of interstellar and intergalactic space, the plasma is so rarefied that a cubic centimeter contains almost nothing—but it turns out that *almost* nothing is enough for sound to exist. Nevertheless, "In space, no one can hear you scream" is true, because there simply isn't enough matter to carry soundwaves that human ears can hear.

Remember that for sound to travel through any medium, its wavelength must be far larger than the "mean free path" which is the average distance traveled by particles in the medium before they bump into another particle. A sound at 440 Hz (the tone defined in Western music as concert A) has a wavelength of 75.9 cm. In the dense air of Earth, it will encounter many billions of molecules it can cause to bump into adjacent molecules within this distance. In space, concert A would have no way of propagating: the mean free path out there is vastly longer than its 75.9 cm wavelength. However, for an ultralow-frequency (ULF) sound with a wavelength measured in light years, even in intergalactic space there will be vast numbers of particles over its enormous wavelength. And plasma as a medium has a significant advantage over other states of matter: its charged particles don't have to make physical contact in order to push or pull another particle; they can do so from a distance. To sounds with sufficiently long wavelengths, the sparse plasma of space is a perfectly workable continuous medium. An eardrum the size of a galaxy with ULF sensitivity would hear a universe alive with sound, including the chiming of stars, the roar of black holes eating the matter around them and the cataclysmic detonations of supernovae.

In the almost fourteen billion years since the Big Bang, the universe has expanded massively, so the plasma has become very diffuse. However, gravity centered on dark matter concentrations has shaped countless regions of space where the density is far greater than average, such as galaxies, stars, planets and the gas clouds we see in the stunning images from the Hubble and James Webb space telescopes. As we've just discovered, contradicting the common misconception that space is silent, sound exists almost everywhere—but in these denser regions, it positively thrives, and some of it would be audible to human ears.

The singing stars

The most numerous sound-makers in the universe are stars, which, scientists have recently discovered, ring like bells.

The massive continuous explosions of heat and light at the heart of a typical star give rise to vast convective motions of plasma, creating pressure waves that make the star ring with a unique set of tones depending on its size, composition and structure. Mark Whittle gave me an analogy with a more familiar object: "A bell has a finite structure and boundaries, and waves move around it, so the bell can be set up in various modes of oscillation. The same is true in a star, and each star actually has its own voice."

The frequencies of the sounds within stars and the resulting oscillations of their surfaces are mainly very low. Large things have lower voices throughout nature, and stars are the largest single objects in the universe—so it's perhaps no surprise that some have vibrational frequencies as low as once every ten years. The Sun's surface pulsates roughly once every five minutes.

As scientists have come to understand that stars sing, a new science has arisen and transformed our understanding of them. It's called asteroseismology (or, in the case of our own star, helioseismology) and it studies the oscillations of stars created by the soundwaves inside them. Just as seismology on Earth has allowed us to analyze the internal composition of our planet right down to its core, so asteroseismology is allowing scientists to work out what is going on inside even the most distant stars.

By shifting the very low frequencies that stars and other objects emit upward until we can hear them, we now know that stars sing multiple notes simultaneously as soundwaves travel through them and bounce back from their surfaces, and that each star has its own song. Larger stars apparently sing fewer notes. Listening to the songs of distant stars by measuring the frequencies of the tiny pulsations in their light allows scientists to deduce a wealth of information about them, including their densities, rotation rates and magnetic fields.

California Institute of Technology professor Jim Fuller is an asteroseismologist who has sonified numerous stars by speeding up their songs by a factor of around a million and has used the analysis of the songs of distant stars to discover many multiple stars that were hitherto thought to be single: binary and multiple stars orbiting one another sing in special types of harmony as their gravities pull on one another and cause fluctuations in their surfaces. Fuller's sonifications show how, while every star has its own unique sonic signature, different types of star have their own classes of song [91].

Very recently, we've learned that there are even soundwaves ringing in the massive coronal loops that spectacularly leap out from the Sun's surface and arc back again. The sensitivity of the detectors and filters now being used on the Sun, and the sophistication of the models that scientists have developed, mean that we can now detect sunspots on the far side of the Sun long before they come into view and potentially cause coronal mass ejections that could be highly damaging or even fatal to astronauts and satellites in space, as well as to power grids and electronics on Earth. Once again, sound opens a new dimension, enriching our understanding of the universe around us and helping us to avoid or at least prepare for disasters.

Sound creates structure

The driving force behind most of the sound in space, of course, is gravity. It may be a weak force, but when gravity acts on gas or plasma over millions of years, causing it to fall faster and faster into a gravity well, this creates pressure waves. These are real soundwaves until they accelerate past the speed of sound, at which point matter can't get out of the way fast enough and, like a broom pushing a pile of earth, a shockwave propagates.

These sound- and shockwaves exist all over the universe, contributing to the forms that we see in the majestic spiral arms of galaxies, as well as magnificent dust clouds like the famous Pillars of Creation, and even the formation of stars. With huge wavelengths up to many light years in length, these ultralow-frequency sounds are far below human hearing, and their energy is mostly too small for our instruments to detect—though when gravity moves gas around, the forces can sometimes be significant and the gas can move at speeds of many thousands of kilometers an hour.

When gas collides at these hypersonic speeds, it produces ionization and immense temperature, which is why, counterintuitively, deep space is very hot: the plasma's temperature in the depths of intergalactic space is in the millions of degrees. However, the particles of matter out there are so sparse that there is no danger of a future interstellar traveler being cooked: the much more populous photons from the Big Bang are very cold now, at just 2.7°K (minus 270°C), which is what you would feel if you unwisely removed your space suit.

So it's strange but true to say that deep space is both noisy and hot—although human perception would not sense either of these things.

Aural astronomy

Not only does sound exist throughout the cosmos, it can also be a dramatic, engaging and beautiful lens through which to better perceive and marvel at the extraordinary objects out there. We can turn visual data, or any form of vibration, into audible sound, revealing hidden trends or forces, creating beautiful and eerie sounds, and excitingly opening the fields of astronomy and cosmology to people with impaired vision or blindness. For these reasons, aural astronomy is a fast-growing and exciting new field. It started in 1987, when

Italian astrophysicist Fiorella Terenzi used pitch-shifting to open up a whole fresh approach to astronomy, now known as "sonification," when she converted radio observations from the galaxy UGC 6697 into sound for her Ph.D thesis. Now an award-winning professor at Florida International University, she produced an internationally successful CD of sonifications called *Music from the Galaxies*, and has collaborated with Herbie Hancock, Ornette Coleman, Massive Attack and Thomas Dolby, among others.

Sonification once turned a comet into a star, metaphorically speaking. In November 2014, the European Space Agency (ESA)'s *Rosetta* probe encountered comet 67P/Churyumov–Gerasimenko and recorded some interesting low-frequency oscillations in the comet's magnetic field. The ESA scientists scaled the vibrations up by a factor of ten thousand to make them audible, and posted the sound file, entitled *A Singing Comet*, on the SoundCloud platform, where over five million people listened to it [92].

Sonifying the cosmos can even create music. Take a photograph from the Hubble space telescope and map visual features onto musical ones—for example, brighter objects are louder, different colors or objects are represented by specific instruments, while different distances generate different notes—and you can listen to the picture as you scan across it [93]. Using mapping like this, there is a beautiful sonification of Hubble's Ultra Deep Field picture from 2014, where celestial objects are successively added to the picture according to their distance from Earth so that the sonification becomes increasingly complex until the entire Deep Field is revealed [94].

The same approach translates images from the James Webb and Chandra space telescopes into mainly musical sound with engaging results that you can enjoy at the SYSTEM Sounds website [95] and the Chandra website [96]. A group of Harvard University scientists and sound engineers have created similar sonifications [97].

Matt Russo is a physics professor at the University of Toronto specializing in the connections between music and astronomy, and is a founder of SYSTEM Sounds, the organization behind many of these lovely sonifications. I spoke to him at his home in Toronto and asked him why aural astronomy was important. His answer: "There's a puzzle that has to be solved in a way inside each listener's head when they listen to a sonification. Maybe that's why it's so engaging, because when NASA posts our work, the numbers are huge. People respond much more than I would have expected to image sonification in particular. The more concrete benefit, though, is making astronomy accessible to people who are blind or visually impaired. That's one of our main goals." And is it growing? "We're undergoing a huge explosion right now in sonification. There's a paper in *Nature* by some colleagues who found that in the last five years it's been exponential. There's also an explosion of tools…so people who are not sound or music experts are able to use programs like TwoTone.io to create an audio version of any dataset."

Aural astronomy can also cross over from science to art. Of course, many composers have written pieces either inspired by the cosmos or featuring space sounds: Holst's *The Planets* is probably the best known, but modern composers have been able to incorporate actual space sonifications into their works. To cite just a few, German electronic composer Robert Schröder released *Galaxie Cygnus-A* in 1982, featuring radio signals from that galaxy [98]; in 2014 *The Space Project* compiled songs by various bands based on recordings from *Voyagers 1* and *2*; while in the same year the collective Fabrica Music Area released *80UA*, a four-track EP featuring sounds from the NASA Sound Archive [99]; Terry Riley's 2002 composition *Sun Rings*, performed by the Kronos Quartet, used a selection of Don Gurnett's space sound recordings; and in 2012 Mickey Hart, the former drummer for the Grateful Dead, teamed up with Nobel Prize-winning cosmologist George Smoot to turn the frequencies of the universe into music, shifting light and electromagnetic waves into the audible range in a twelve-track work called *Mysterium Tremendum* with an engaging film called *Mickey Hart's Rhythmic Universe* [100].

As Matt Russo says, one very important benefit of aural astronomy is that it makes the subject accessible to people with impaired vision—like the blind Puerto Rican astronomer Wanda Díaz-Merced, who lost her sight in her early twenties due to diabetes. When she heard of NASA's Radio JOVE (a project where students and amateur scientists from around the world observe and analyze natural radio emissions of Jupiter, the Sun and our galaxy using their own easy-to-construct radio telescopes), she realized that her passion for science did not have to be halted by her blindness. In the years since then, she has become a leading evangelist for the use of sound in astronomy, and delivered an inspiring talk at TED2016 [101]. She is now taking this new way of experiencing the cosmos to thousands of citizen scientists through the Research Infrastructures for Citizens in Europe (REINFORCE) program, using sonification to help them seek new insights in the mass of data that's being collected all the time by various instruments in radio, x-rays and light.

Sonification is an approach that has huge potential in many fields, not just astronomy. It's already used in some financial trading, where dealers can add audio signals and alarms to the bank of screens surrounding them with visual data. We are all familiar with the auralisation of heartbeats in hospital monitors, but there are countless data streams in healthcare or environmental analysis that can just as easily be turned into sound for better understanding and faster response by humans. This is a rich resource for future exploration.

Magnetic music

Astral bell-ringing is not the only way stars create sound. The flows and currents inside our Sun that create its song also generate enormously powerful magnetic fields, which in turn create something we call the solar wind. This is a stream of charged particles rushing out into space in all directions at over a million kilometers an hour, as even the Sun's powerful gravity can't contain all of its super-heated plasma. The solar wind is very variable, especially when huge surface explosions, known as solar flares and coronal mass ejections, happen.

This roaring solar wind batters all the planets. Some, including the Earth, have a defensive shield called a magnetosphere—a powerful magnetic field that deflects away most of the torrent of charged particles from the Sun, as well as stray cosmic radiation from deep space, making our existence possible. But the magnetosphere is not impregnable. Strong surges of solar wind can crack or even penetrate it, causing "solar weather" that, at the extreme, can disrupt electrical, radio and electronic networks, destroy satellites and harm astronauts.

As the magnetosphere reacts to the powerful impacts of the solar wind, scientists have now discovered that it acts like a huge musical instrument and makes real sounds. Martin Archer is the UK Research and Innovation Stephen Hawking Fellow in Space Plasma Physics at Imperial College, London, as well as a DJ and popular science broadcaster. In a call with me, he described how the solar wind creates ultralow-frequency soundwaves that move through the plasma around our planet. As usual with genuine sound in space, these soundwaves have very long wavelengths, because the plasma is so sparse and the mean free path very long. This is genuine sound, but far below the range of human hearing. "The waves I look at are at milliHertz frequencies, so in order to hear them we speed them up by tens of thousands of times," Archer said, explaining to me that there are three kinds of magnetosonic sound out there.

First, buffeting from the turbulent solar wind causes the edge of our magnetosphere to resonate like a drum being struck by a stick. This booming sound has been theorized for decades and was confirmed recently by observations from new satellites.

Second, the Earth's magnetic field lines behave like a set of guitar strings, each tuned to a distinct note: the solar wind can set them in motion so that they create tones and harmonics. Archer told me, "Every field line will have a different natural frequency it wants to ring at, and we often find that they are ringing at these frequencies."

And, finally, there are soundwaves bouncing between the boundaries of different layers in the magnetosphere, which act like the resonating chamber of a violin.

The whole thing produces a complex soundscape that takes some decoding. In one of his educational videos, Archer summarizes: "The magnetosphere is probably one of the most complicated musical instruments that exists. We're trying to pull apart a symphony of different notes that have been played in different ways to uncover the mystery of not only the player but the composer as well."

Archer has filmed a good introduction to space sound, explaining plasma, electromagnetic and gravitational waves, with audio examples included [102].

Earth is not the only body in the solar system to have a vast magnetosonic instrument: Mercury, Jupiter, Saturn, Uranus and Neptune, as well as Jupiter's largest moon, Ganymede, also have their own. Mars, Venus and Pluto do not, though their ionospheres can still sing in response to the solar wind—and this is also true of comets. From observations by the spacecraft that we've now sent to every significant body in the solar system, it seems they are all singing, and the songs are unique to each body. Even without these interactions, the Sun is broadcasting its own magnetosonic voice. As Archer told me, "Magnetosonic waves can be also be present in the solar wind itself. They can just propagate out from the Sun and go all the way out to the edge of the solar system."

Fascinatingly, we now also know that magnetospheres are not confined to our solar system. The prosaically named exoplanet HAT-P-11b, discovered in 2009, is a "hot Neptune" planet orbiting close to a star around 123 light years from Earth. In December 2021 an international team of scientists using data from the Hubble space telescope established that HAT-P-11b has a magnetosphere, so it's likely that the space within every solar system in the universe is alive with sound, all ultimately stemming from those singing bells, the stars.

Catastrophic sound

Moving gas around is not the only way in which gravity creates real soundwaves in deep space. The same gentle force, some 1040 times weaker than the electromagnetic force that binds atoms together, also creates cataclysmic events that generate enormous soundwaves.

According to the U.S. Department of Energy, a star explodes every ten seconds somewhere in the known universe, while the European Space Agency's research shows that there is a supernova roughly once every fifty years in our own Milky Way galaxy. These titanic explosions create colossal shock- and soundwaves that travel through surrounding gas, dust and plasma.

Most supernovae happen when an enormous star runs out of fuel for atomic fusion. When that happens, the outward pressure of the fusion explosion can no longer push back against the star's massive gravity, so the entire structure suddenly collapses inward, creating a huge explosion. These are some of the loudest events in the universe. When such a supernova happens, an imaginary being with vast ears and sensitivity to ultralow-frequency sound would hear a tremendous explosion as shockwaves rip through the interstellar plasma at several percent of the speed of light.

We are all children of supernovae. Aside from hydrogen, every atom in our entire planet—including the ones inside you and me—was forged in the center of a star that died in one of these cataclysmic detonations billions of years ago. The stars are the crucible of all cosmic chemistry, forging complex atoms from simpler ones, and supernovae are so intense that they take the whole process up several notches, creating heavy elements that are beyond even the fusion in the center of a star to create, and then blasting them out into clouds of gas and dust that eventually coalesce to form new solar systems.

A supernova is not the end of the process for a dying star, and the frankly unimaginable next stage can itself create powerful waves in space. What's usually left after a supernova blows off the bulk of a giant star is what astronomers call a "compact object"—a neutron star, or possibly a black hole if the original star was vast enough.

These are some of the strangest objects in the universe...and we can listen to them.

A neutron star is essentially a giant atomic nucleus. The enormous gravitational pressure fuses together all the protons and electrons in the collapsing star's core, cancelling their positive and negative charges and turning them into neutrons. When the process ends, what's left is a sphere roughly the diameter of London or New York, but with the mass of the Sun. A teaspoonful of this star would weigh about a hundred million tons, and a human standing on the surface would be instantly crushed into a paste just a few atoms thick. If that weren't enough, the star is spinning at incredible speed. Just as ice skaters accelerate their spinning speed by drawing in their arms and becoming more compact, so the original star's rotation speed gets hugely boosted as its diameter drastically reduces.

Neutron stars have magnetic fields around a trillion times as strong as Earth's, and they emit high-energy beams from their magnetic poles. If the beams are pointing toward our radio telescopes, we can detect the rotation as pulses, hence the name "pulsar" given to these objects. The frequency of the pulse is often in the audible spectrum: the record holder to date is the catchily named PSR J1748–2446ad, some 18,000 light years from Earth in the Terzan 5 globular cluster in the constellation Sagittarius, which is spinning an incredible 716 times every second. That's a very audible frequency: 716 Hz is halfway between the F and F sharp above concert A. Of course, these are electromagnetic waves pulsing, not actual sound: here again, we're sonifying something to gain a new way of experiencing it. You can hear various pulsar sonifications at the Jodrell Bank website [103].

Now try to imagine an object the size of Manhattan with the mass of the Sun spinning at over seven hundred times a second. The surface at the equator of this extraordinary object is spinning at almost a quarter the speed of light, and its shape is distorted by a significant centrifugal bulge. Humanity may never actually behold such a thing, but, remarkably, from the vibrations alone, scientists have been able to deduce much about these prodigious objects.

There are even bigger forces out there, and they create soundwaves so powerful that astronomers on Earth have detected them. The Perseus cluster is one of the largest structures in the known universe, containing thousands of galaxies surrounded by a superheated cloud of gas; at its center lies a supermassive black hole. In 2003 a team from Cambridge University analyzed fifty-three hours of readings from the Chandra X-ray Observatory space telescope and detected powerful soundwaves generated by the black hole causing huge ripples in the plasma of the local space. They calculated the sound's frequency as once every 9.6 million years, which is a note some fifty-seven octaves below middle C. NASA has created an eerie but not inharmonious sonification by pitch-shifting the sound up to the audible range [104].

It seems likely that all black holes create sound. At the heart of galaxy M87 in the Virgo cluster lurks one of the largest black holes yet detected, now famous after it was photographed by the Event Horizon telescope, and this, too, has been shown to be a powerful sound-maker. Its tones are more discordant than the Perseus black hole, according to the team from the Harvard–Smithsonian Center for Astrophysics (CfA) who analyzed the data, again from Chandra, in 2006. "We can tell that many deep and different sounds have been rumbling through this cluster for most of the lifetime of the universe," commented the CfA's William Forman.

There is one more kind of cosmological wave we can now detect and sonify. If two compact objects—two neutron stars, two black holes, or a neutron star and a black hole—form a binary system, they will inevitably eventually smash together in an event so powerful that it will create ripples in the fabric of space itself. These are known as gravitational waves, and the event produces a distinctive waveform known as a "chirp"—an innocuous name for something emanating from one of the most cataclysmic events in the universe.

Mark Whittle described to me this accelerating dance of death:

> The gravity of that system is so powerful that it begins to radiate gravitational waves... space itself is wiggling, being radiated away at the speed of light, and that drains energy from the orbiting objects. If you take energy out of two orbiting objects, they get closer together and they start speeding up faster, and that makes even more powerful waves and that takes more energy away, so you end up with this rapid runaway process whereby the objects get closer and closer.
>
> Now let's just put some numbers in here. Each black hole might have a Schwarzschild radius* of say three kilometers and they might be a thousand kilometers apart, orbiting maybe once every second. So you get a gravitational wave moving out with a period of one second, but gradually these things get closer and closer until they're zipping around each other at thirty times a second, forty times a second, a hundred times a second. And finally their event horizons merge in a much more tumultuous merger process and they become a single black hole, which then settles down quite quickly—it takes just a few milliseconds. And

---

* The radius of the black hole's event horizon.

that whole process has pumped out gravitational waves, and that's what generates these very characteristic chirps.

There's actually quite a complex and extremely interesting waveform at the very end. And if you're able to track that and measure it, it allows you to figure out the spins of the black holes, the orbit shape, their relative masses, all sorts of things. It's a unique fingerprint of the event.

We now have instruments sensitive enough to detect the infinitesimally faint gravitational waves that reach Earth from such mind-blowing events, all of which are fortunately very distant. The first ever observation, named GW150914, was made by the Advanced Laser Interferometer Gravitational-Wave Observatory (LIGO) instruments in Hanford, Washington State and Livingston, Louisiana, in collaboration with the Virgo instrument near Pisa, Italy, on 14 September 2015. From the waveform, the scientists worked out that what they had detected were the gravitational ripples resulting from the merger of two black holes. The frequency of the vibrations was already in the audible range, but pitch-shifting them up by a little over an octave makes it easier to hear the chirp recorded by these impressive instruments, whose arms move by just a hundredth of the diameter of a proton when gravity waves pass by. LIGO has made a short film explaining gravity waves and how they are measured, with audio from this first measurement included [105].

Since that first breakthrough, the LIGO network has detected several more of these faint ripples in the fabric of spacetime here on Earth that speak of distant cataclysmic events, including at least one resulting from the collision of two neutron stars.

Sound created from waves in spacetime itself is perhaps the

ultimate in cosmophony, but there is plenty of sound in space that's happening much closer to us.

Alien weather

Based on our current knowledge, the probability of multicellular life elsewhere in our solar system is very low, so Earth is probably the only place a listener can experience biophony; unless, and until, we establish colonies on the Moon or Mars, it is certainly the only place with anthropophony. However, we now know that there is plenty of geophony-like sound on alien planets and even moons. A sonic tour of the solar system will reveal to us some surprising places that have soundscapes we would hear by positioning an imaginary, indestructible microphone there—although some might need amplifying to become audible. We'll start with our neighbors, the rocky inner planets.

Any planetary atmosphere is a medium that can carry soundwaves, and we now have audio recordings of two of our three nearest neighbors to prove that extraterrestrial sound is a reality. The exception is Mercury, which has almost no atmosphere at all. The only sound a human would hear there (if they could survive the somewhat extreme climate, which seesaws between 400°C in the daytime and minus 180°C at night) would be seismic movements (Mercury-quakes) heard by pressing an ear to the scorching or freezing rock, which is probably more effort than would be justified by the result.

However, as we discovered earlier in this book, the Earth resonates with what are called free oscillations (the hum) that scientists theorize are set off by seismic activity and deep-ocean waves, or, even more intriguingly, by some unknown inner process. It's likely that all the rocky inner planets do this, and it's reasonable to speculate that every planet will have its own fundamental resonant frequency, possibly with harmonics forming a unique sonic signature as a very low-frequency chord.

The only other planet to hint at possessing inner sound is Saturn: its pulsations have a discernible effect on its rings, which act as a giant seismograph, amplifying the movement just as a stylus amplifies the grooves of a vinyl record. Not content with analyzing the songs of distant stars, Jim Fuller has used similar techniques to take Saturn's pulse. He says, "The pulsation causes a spiral density wave in the rings that propagates at the same rate. The motion of these spiral patterns then tells us the frequency at which Saturn pulsates, and this can be used to measure properties of the interior of Saturn. My work suggests that Saturn's interior is more complex than previously believed, with an outer core composed of a mix of the icy/rocky core material and the gaseous material found in the outer envelope."

Perhaps one day we'll have instruments sensitive enough to detect the signature sounds of all the planets, which will probably reveal a great deal about their structure and the history of both the solar system and the planets themselves, as well as yielding fascinating sounds to listen to.

Leaving Mercury behind, we come to Earth's tortured twin, Venus. Driven by a runaway greenhouse effect, Venus's temperature at ground level is even higher than Mercury's, at almost 500°C, making it the hottest planet in the entire solar system. This is not a place for a visit. The surface pressure is around ninety Earth atmospheres (the equivalent of being nine hundred meters deep in the sea on Earth, enough to crush you or me to death instantly) and the air is a noxious mix of carbon dioxide and sulfuric acid. At high altitudes 300 kph winds whip the clouds around, so there is plenty of wind noise up there, along with the sound of sulfuric acid rain hitting our imaginary microphone. On the surface, the viscous air moves far more slowly, while any rain evaporates in the intense heat long before reaching the ground, so it's likely that at ground level the dominant sound will be thunder, as lightning strikes pierce the yellowish gloom. The only other sound might come from an eruption of one of the volcanoes pockmarking the planet's surface, or from a Venusquake. Added to all the other hellish aspects of this once-habitable planet, the soundscape on the surface is thus probably sullen silence punctuated by threatening, explosive noises.

Venus was the location of the very first audio recording sent back from another planet. In 1982 the *Soviet Venera 13* and *14* probes successfully landed on the Venusian surface, and they both had microphones whose primary aim was to assess wind speed. The *Venera 14* audio recording is publicly available and is a staggering achievement given the technology available at the time. In what is probably the most challenging sound recording ever made by humanity [106], we can hear the sound of Venusian breeze (scalding carbon dioxide at immense pressure grinding past the mic at around 3 kph), punctuated by noises from the probe as it urgently sets about its work, knowing that its life expectancy was probably just minutes. First, there are pops as the camera lens cap is blown off, then *Venera 14*'s drill operates to take a sample of Venusian soil, ending with a dull thud as the soil sample lands in the analysis vessel. The plucky probe actually survived for around an hour before succumbing to the infernal conditions. Try to visualize the context of this recording as you listen, and your emotional response may well be awe.

NASA's *Perseverance* rover is probably still trundling across the surface of Mars as you read this, tirelessly seeking any evidence of life, whether microbial or more complex. Among its mass of experiments, robotic limbs and sensors, *Perseverance* carries two ordinary shop-bought microphones that have been capturing the sound of Mars, and the rover has been diligently sending the audio files back for us to listen to. The atmosphere on Mars is over a hundred times thinner than our own and comprises mainly carbon dioxide. As a result, sound propagates differently there, feeling muffled and lacking higher frequencies, and it travels more slowly, at around 250 meters per second, compared to around 330 mps on Earth.

I find it thrilling to listen to the recording *Perseverance* made on its fourth sol (Martian day)—of a gentle breeze on Mars. It's such a familiar and pleasant sound, and yet so chillingly unfamiliar when you imagine the lone rover recording it on this freezing cold, barren planet, eighty-five million kilometers away [107].

Mars has virtually no precipitation and, as far as we know, the planet is geologically inert, its many huge volcanoes all extinct, so the surface is very quiet unless you are standing right by one of the carbon dioxide geysers that scientists believe erupt frequently from the ice in the South polar region of the planet—or if you happen to be in a dust storm. Perhaps one day *Perseverance* will record the 100 kph winds of such a storm. That wind speed sounds intimidating, but the air is so thin that it would have very little physical impact. Martian dust storms may be enormous, sometimes covering the entire planet, but they are relatively mild events by solar system standards.

The average distance of Earth from the Sun is around 150 million kilometers, which astronomers designate as one astronomical unit (1 AU). All four rocky inner planets orbit within 1.5 AU of the Sun.

After Mars, the distances between planets become enormous, and so do the planets themselves. Jupiter is around 3.7 AU past Mars, Saturn another 4.3 AU past Jupiter, and then it's around 10 AU to Uranus, and the same again to Neptune.

We have no audio recordings from the four giant planets, but there is one astounding piece of sound recorded on one of their moons. On Christmas Day 2004, in the icy depths of space, over a billion kilometers from its control center on Earth, the *Cassini* Saturn orbiter ejected *Huygens*, a robotic probe, toward Saturn's largest moon, Titan. This little probe was carrying microphones.

Almost half the size of Earth, Titan is the only moon in the solar system known to have its own atmosphere, which is almost entirely composed of nitrogen. It is also the only body we know of (apart from Earth) that has liquid on its surface in the form of lakes, rivers and rainfall. However, at minus 180°C this liquid is not water (which exists on Titan only as a solid that forms rocks and mountains), but methane.

On 14 January 2005 *Huygens* touched down on the surface of Titan, in doing so achieving by far the most distant spacecraft landing that humanity has accomplished to date. During the two-and-a-half-hour descent through 400 kph winds in the upper atmosphere, its microphones were recording, and NASA has made a short edit of that recording available. It's another astounding piece of sound and it remains the most distant audio recording ever made [108].

Huygens took photographs and sent back data from its various instruments for around ninety minutes before it died. Because of its distance from the Sun and the thick haze of its cloudy atmosphere, the surface of Titan receives only around a hundredth of the sunlight that Earth does. The light is comparable to Earth ten minutes after sunset—the predominant color orange. The famous Huygens pebble picture, which speaks of terrain shaped by running liquid, shows a quiet-looking plain that may have once been the bed of a lake. Although there is no sound recording from the surface, we do know that, as on Mars and Venus, winds on the surface are mainly gentle, although *Cassini* has witnessed dust storms, monsoon-like rains and stronger seasonal winds that ripple the surface of the methane lakes.

Sound would carry well in Titan's atmosphere, so this distant moon has probably the most Earth-like soundscape in the solar system, featuring wind, rain and running liquid. However, from time to time extraordinary explosions might interrupt this familiar soundscape as cryovolcanoes or cryogeysers erupt and eject liquid water, ammonia or methane that freezes instantly on contact with the atmosphere.

Similar eruptions exist on Saturn's other moon Enceladus, as well as Jupiter's moons Europa and Ganymede, though since these moons have no atmosphere there would be no sound from the cryoeruptions. However, a different kind of sound almost certainly does exist on these three moons, which are possibly the most intriguing bodies in the solar system.

Saturn's moon Enceladus is just 500 kilometers in diameter. Its surface is solid ice, and it was long thought to be an inert, frozen ball, until Cassini discovered more than a hundred cryogeysers shooting liquid water far out into space—a startling discovery of a process now known to be the source of Saturn's icy E ring. Scientists believe this water comes from a huge subsurface ocean of warm liquid water underneath a forty-kilometer-thick shell of ice—an ocean that is around ten kilometers deep, and is heated from within the moon by tidal gravity. In 2015 scientists analyzed silicate particles in the E ring and were able to deduce that there is geothermal activity taking place deep in Enceladus's under-ice ocean, just like the hydrothermal vents in the ocean deeps on Earth. These are the exact places where many people believe geochemistry became biochemistry, giving rise to life on Earth some three billion years ago.

Enceladus is not the only moon with a hidden surprise. Six times as large, Jupiter's moon Ganymede is thought to have a very similar structure, with an ice crust on top of a vast, 800-km-deep subsurface ocean. And not far away from Ganymede is Europa, much smaller but equally believed to be hiding a subsurface ocean that's up to a hundred kilometers deep which, according to data from the *Galileo* probe, contains three times as much water as all Earth's oceans combined. There is evidence of cryogeyser activity on both moons, albeit less strong than on Enceladus.

The high probability of hydrothermal vents on the floor of all three of these subsurface oceans makes them leading candidates in the search for extraterrestrial life, whether microbial or multicellular. But even if these dark oceans are in fact lifeless, they almost certainly contain sound. We know how well water conducts soundwaves, and hydrothermal vents on Earth have been recorded and measured producing sound at up to 30 dB above the ambient noise level [109].

There are numerous expeditions proposed to all these moons in the decades to come, aiming to scan them from close range and sample the water being ejected by their cryogeysers. The prize is the revelation of a second, independent bloom of life, made all the more enticing by the discovery in 2014 of life in Lake Vostok, a kilometer-deep body of water that exists beneath three kilometers of ice in the Antarctic. NASA has been researching probes that could bore through kilometers of ice and survive the enormous pressures in the oceans below, but the technical challenges of doing that a billion kilometers distant are mind-boggling, even compared to surviving the nightmare conditions on Venus, so sadly I don't expect to hear a recording of sound from any of these mysterious oceans in my lifetime.

Other moons in the solar system have atmospheres, but they are so thin that there's no weather to speak of and human ears would detect no sound at all—even from the hundreds of active volcanoes on Jupiter's pockmarked moon Io—unless pressed to the ground to sense seismic waves through bone conduction.

And so we come to the solar system's four enormous outer planets: the gas giants, Jupiter and Saturn, mainly composed of hydrogen in gas, liquid and metallic fluid forms; and the ice giants, Uranus and Neptune, with their blue methane clouds and icy centers. Together, they account for 99.5 percent of the planetary mass in the solar system. All four have plenty of sound that our imaginary microphone could record, arising from violent, and sometimes fantastical, weather in their dense atmospheres.

First, there will be thunder. *Voyager 1* detected lightning as it flew past Jupiter in 1979, then the *New Horizons* and *Juno* probes measured strikes ten times as powerful as anything on Earth. As *Voyager 1* moved on past Saturn, it too detected sferics (microwave-frequency signals made by lightning); *Cassini* later established that Saturn's lightning happens up to ten times per second in powerful storms. *Voyager 2* detected whistlers (radio emissions made by lightning) coming from both Uranus and Neptune. Lightning creates thunder, so the storms on these strange worlds are noisy events—and thunder is only part of the racket. The main noise producer will be the wind.

On Earth, winds of 400 kph occur occasionally in hurricanes

or tornadoes. On Neptune, *average* wind speeds are faster than the speed of sound (1,200 kph), and in the storm of its Great Dark Spot winds of over 2,000 kph have been recorded—the fastest in the solar system. Here in Orkney, we're very familiar with wind chill in our strong winter gales, but the mind boggles at the effect of 2,000 kph wind speeds on methane clouds that are at minus 200°C to start with.

The other giants are not much less violent in their weather. Saturn has winds exceeding 1,500 kph, while Jupiter's Great Red Spot is a single storm four times the diameter of Earth that has raged for over a hundred years and contains winds of over 600 kph. Recent analysis by the European Southern Observatory using the Atacama Large Millimeter/submillimeter Array has found winds of up to 1,450 kph near Jupiter's poles. Uranus is not far behind: its frozen upper-atmosphere winds often exceed 900 kph.

All these planets have such deep atmospheres that the pressure low down becomes millions of times greater than Earth's sea-level air pressure (one atmosphere). If we could descend to the small rocky core that's believed to be at the very center of Jupiter, the temperature would be over 20,000°C and the pressure some fifty million atmospheres. This immense pressure in the atmospheres of all four giant planets is believed to give rise to a very exotic type of precipitation that would certainly make some sound as it rattled off our imaginary microphone: billions of diamonds, each around a centimeter in diameter, falling as priceless rain.

This is how it happens. As lightning flashes in the upper atmosphere of all four giant planets, it transforms methane in their clouds into carbon, in the form of soot. The soot then falls, encountering ever-increasing pressure. First it's squashed into clumps of graphite, then as these fall further, the intense pressure compresses the clumps into diamonds. On Jupiter and Saturn, even diamonds can't survive the extreme pressure and heat in their lower atmospheres: scientists are uncertain what happens next, but theorize that the gems may melt into liquid carbon. However, on the much colder Uranus and Neptune, the diamonds may well survive, floating in the liquid mantle or possibly vaporizing and floating back up through convection to form yet more diamond rain.

We now also know that weather is not confined to our solar system. In 2015 astrophysicists at the University of Warwick analyzed the exoplanet HD 189733b, a "hot Jupiter" around sixty-five light years from Earth in the constellation of Vulpecula, and found that its winds reach almost 9,000 kph, and that its version of precipitation is molten glass, moving horizontally at vast speed in its 700°C atmosphere. The roar of this deadly weather is beyond imagining.

Although our probes have now visited every known planet, as well as several planetoids and asteroids, much of what we have discovered simply poses new and intriguing questions: for example, the quick fly-by of Pluto by the *New Horizons* probe intriguingly suggested that this distant, freezing planetoid may have an internal heat source. In years to come, many more wonders will be revealed, and perhaps unfamiliar sounds will be key to those discoveries.

The elements of the soundscapes on these profoundly alien worlds may be familiar—wind, thunder, rain and hail, volcanoes, seismic events and subsea vents—but the scale and sheer violence of most of them are without precedent on Earth. Sound in these places must be savage, oppressive and incessant. I find that trying to imagine the brutal soundscapes that exist on almost the entire planetary mass of the solar system is a fine way of accessing deep gratitude for the myriad exquisite, gentle sounds we can enjoy every day on our small blue dot.

Listening to electromagnetism

Some electromagnetic radiation doesn't need to be pitch-shifted because it's already in the range of human hearing. This kind of sound was first detected during the First World War by a German engineer called Heinrich Barkhausen, who was trying to intercept British field telephone calls, using long wires connected to a high-gain amplifier. What he heard instead were strange high-pitched descending sounds now known as "whistlers." It wasn't until 1953 that physicist Owen Storey established these sounds were caused by lightning strikes sending plasma waves upward. Each wave travels along one of the magnetic field lines that radiate out from and then circle back to the surface of the Earth. Since high frequencies travel faster through plasma than low ones, a receiver at the point where the field line rejoins the Earth will hear the soundwave as a short, rapidly descending tone—hence the name whistler.

Whistlers are rare to detect on the ground, but when scientists put antennae on satellites in the early 1960s they heard a cacophony of them, along with other plasma-borne waves emanating from Earth. As a result, starting with *Voyager 1*, interplanetary probes have been fitted with detectors that allow them to listen for these sounds.

For over fifty years, until his death in 2022, the University of Iowa's professor Don Gurnett researched, published papers on, and collected recordings of space sounds like these. He was involved in over thirty spacecraft projects, including both *Voyagers*, *Galileo* and *Cassini*.

The instruments that he helped to design and create collected (and in the case of *Voyager 2* still collect) many audio samples, and the university maintains an excellent website called Space Audio, which hosts a fascinating selection of them, with detailed explanations [110]. The site has many plasma-borne sounds to enjoy, including whistlers and chirps from both Earth and Jupiter, as well as eerie sounds emanating from the Sun, recorded in interstellar space by *Voyager 1*. All these sounds are in the kilohertz range and therefore audible to our ears, though they are generated electromagnetically rather than mechanically: they are plasma waves driven by electrical fields. In a talk preceding a performance of Terry Riley's *Sun Rings*, which used a selection of his space sounds, Don Gurnett said, "Are these really soundwaves? I think if we could make a microphone sensitive enough, we could detect them in space—though as they have an electrical character, it's easier to detect them with an antenna than with a microphone."

Perhaps the centrality and ubiquity of sound in our universe shouldn't come as a surprise. After all, humans have intuited the importance of sound in explaining the universe for thousands of years.

The universe is sound

Nada Brahma is a Sanskrit expression with roots in Indian Vedic spirituality. *Nada* means sound, while *Brahma* is one of the three core deities in the Hindu pantheon—the creator of the universe, and also its personification. The phrase can be translated in many ways, including "sound is God" (or "God is sound"), "sound is joy," "sound is emptiness," "sound is the fundamental principle," and "sound is the universe"—or "the universe is sound."

Hinduism is not the only spiritual tradition to put sound at the heart of creation. The Old Testament has the heavens and the Earth formless, empty and dark with the spirit of God hovering (alternative translation: vibrating) over them—and only then does God say, "Let there be light." The New Testament says, "In the beginning was the word." The mystics of Islam, the Sufis, say that all form manifests from sound. Various traditions, including Tibetan Buddhist, Aztec, Inuit, Persian, Indian, Malayan, Ancient Egyptian, Polynesian, Japanese, Chinese, Balinese, Tibetan and Ancient Greek have the universe being spoken, sung or chanted into existence, or consider sound (often music) as a primal element in the transformation of a very different starting universe into one that could become home to human beings.

The harmony of the spheres

There is also the millennia-old concept of eternal celestial music, which connects sound, mathematics and astronomy. The theory originated in Ancient Greece and became a core part of Pythagoreanism, and it was developed by the great early seventeenth-century German astronomer Johannes Kepler, who first derived the laws of planetary motion and inspired Newton in his ground-breaking discoveries.

Working with the aristocratic Danish astronomer Tycho Brahe in Prague, Kepler espoused the still-controversial revelation by Copernicus (from only a few decades earlier) that the planets revolved around the Sun, rather than the Earth being the center of the universe—as the Alexandrine astronomer Ptolemy had proposed in his *Almagest* over 1,600 years previously; a theory that had persisted through the centuries. Kepler also challenged two more prevailing views: that the planets and stars were fixed on rotating celestial spheres, and that these spheres were eternal and unchanging.

Based on careful observations of Mars at Brahe's observatory, Kepler instead proposed that there was a force generated by large bodies (he called it "motive power") that inspired the motion of small bodies around them. He later proved that these orbits were ellipses, and between 1609 and 1619 he published what were centuries later formalized by Voltaire and others as the three laws of planetary motion. Some seventy years after Kepler's publications, Newton was able to show that these laws were special cases of his law of gravitation.

In the late sixteenth century, astronomy and astrology were equally respected and often studied together. Kepler was highly religious (although his Lutheran faith caused him many problems in the largely Catholic society he inhabited, and surely contributed to his works being placed on the *Index Librorum Prohibitorum* banned list by the Catholic Church) but he treated astrology as an equal counterpart to astronomy; he spent much time trying to make it more scientific, and horoscopes were his primary source of income for parts of his life. Kepler believed the Earth had a sentient, though not conscious, soul—a view not far removed from James Lovelock's modern-day Gaia hypothesis. He theorized that this was true of all heavenly bodies, and that the interaction between these celestial souls and human ones was the cause of astrological influence on human affairs.

In 1599, in part inspired by Ptolemy's *Harmonikon* and fueled by his deep desire for harmony in a world striven by political and religious conflict, Kepler began working on the *Harmonices Mundi Libri V* (*The Five Books of the Harmony of the World*), his inquiry into the relationship between music, mathematics and the cosmos. Harmony in the broadest sense was Kepler's holy grail, and to him an expression of the divine. In 1605, just after discovering the elliptical nature of Mars's orbit, he wrote in a letter to the astrologer Christopher Heydon in England: "If only God would set me free from astronomy so that I might turn to the care of my work on the harmony of the world."

Published in 1619, the book took the popular medieval metaphor of the music of the spheres (which itself originated in Ancient Greece with Pythagoras) and sought to prove its existence in physical reality. Kepler hypothesized that the motions of the planets would be based on harmonic intervals, because he believed that God's design for the entire universe would be based on harmony.

Kepler assigned a tone to each planet based on its distance from the Sun. Saturn, the farthest planet then observable, had the lowest, with the tones rising all the way to Mercury, eight octaves higher. But each planet's tone was not a constant: it moved smoothly up and down as the planet's angular velocity around the Sun varied—something that's true of all the planets because, as Kepler crucially revealed, the orbits are not circles but ellipses, so the planets speed up when they are closest to the Sun and slow down when they are farthest away. It was in these variations of angular velocity that Keller discovered his harmonies. He found that the ratio of the maximum to minimum velocity for Saturn was 5:4, which in musical terms is a major third. Jupiter had a 6:5 ratio—a minor third. Mars was 3:2, a perfect fifth; Venus almost unity; but Mercury hugely variable at 12:5—a minor tenth. Earth's ratio was 16:15, which is a semitone. "The Earth sings

Mi, Fa, Mi," Kepler wrote. "You may infer even from the syllables that in this our home misery (Mi) and famine (Fa) hold sway."

Kepler discerned more harmonic relationships in the consonant harmonic intervals that arose from the ratios between the average angular velocities of neighboring planets. One clear exception perplexed him: the ratio between Mars and Jupiter was dissonant. Later commentators have speculated that this is because Kepler could not see the asteroid belt between them, with its gravitational effect distorting the pattern.

Kepler saw the planets as a celestial choir, with two basses (Saturn and Jupiter), a tenor (Mars), two altos (Venus and Earth) and a soprano (Mercury). He proposed the planets had sung together in perfect harmony at the moment of creation and might never repeat that until the end of days. He was ahead of his time in taking a positive view of some unconventional intervals and tunings, which would certainly not have been considered harmonious by the musical ears of his time: the planets kept to no well-tempered scales, but varied their tones like a synthesizer moving smoothly through and past notes with continuous glissando, and some of the intervals and harmonies created are decidedly dissonant.

Was Kepler right to see the unseen hand of harmonious design in the motion of the planets? Yes, and no.

Yes, because we now know that gravity does seem to enjoy creating resonant orbits that work in whole numbers. We met Matt Russo a little earlier. In his engaging TEDxUofT talk [111], he played sonifications of two recently discovered exoplanetary systems where the planets do orbit in harmony. Kepler would have been delighted to know of TRAPPIST-1, just forty-one light years from us, where seven Earth-sized planets orbit the so-named red dwarf star in a resonant chain: as the outermost planet orbits twice, the next in goes round three times, then the next four, then six, nine, fifteen and twenty-four—permanently locked. And the star K2-138 (the K stands for Kepler), six hundred light years away, has five planets discovered by citizen scientists in 2017 that were formed with exact 3:2 resonances—a chain of perfect fifths, which Pythagoras would have recognized well. The European Southern Observatory has posted online a pleasing musical animation of the planetary system of a third star, TOI-178, 205 light years from Earth, whose five outer planets are locked in a chain of Laplace resonances* [112].

But no, Kepler was not right, because our solar system does not really sing in harmony: however you map motions onto music, it's quite a discordant set of notes, with rare exceptions like sets of asteroids which are in 3:2, 4:3 or 1:1 resonant orbits with Jupiter, or the 1:2:4 resonance between Jupiter's moons Ganymede, Europa and Io.

However, Kepler would have been thrilled by the deep beauty and mystery that continues to be revealed in the profound relationship between mathematics and the cosmos, from the bizarre contradictions of quantum physics to the grand sweep of Einstein's General Theory of Relativity. The universe is not random or chaotic: there is order and harmony underpinning it at every level, expressed in mathematics—and number forms a bridge from the cosmos to music and audible harmony. Kepler was among the last influential thinkers who could integrate and be open to spiritual, mystical and scientific concepts; for him, harmony was a multidimensional fundamental

---

* A Laplace resonance is a three-body resonance with a 1:2:4 orbital period ratio (equivalent to a 4:2:1 ratio of orbits).

principle, affecting the cosmos, music, human relations and behavior and also our bodies and fortunes equally powerfully.

So much is still unknown to us. Perhaps one day we'll hear the answers rather than see them.

WHAT YOU CAN DO

Contemplating the astounding and often literally unimaginable cosmic sounds we have explored in this chapter is a great way to restore some humility to our often self-absorbed and exploitative relationship with nature—and to regain a sense of how precious and fragile our existence on this planet is. It's well worth setting a little time aside every day to achieve this perspective shift by remembering how very small we really are.

9

SILENCE

> *First, silence makes us pilgrims.*
> *Secondly, silence guards the fire within.*
> *Thirdly, silence teaches us to speak.*
> Henri J. M. Nouwen, *The Way of the Heart*, copyright © 1981 by Henri J. M. Nouwen

Let us return from the depths of space to the place where we began this story: the anechoic chamber, which is as close to the absolute zero of silence as humanity has come—though as we now know, perhaps nowhere in our universe is truly silent. Over the event horizon of a black hole, perhaps silence might exist. Maybe in the heat death of the universe, many trillions of years from now, the soup of photons will be completely still and there will be no pressure waves at all. Perhaps perfect silence is only possible in the complete absence of all matter and energy.

That's rather uncompromising, so let's loosen our definition of silence to mean not this dark and frightening nothingness but instead redraft it as tranquility, or the absence of noise—stillness, quietness and its natural bedfellow, peace. In this softer context, it becomes possible to understand why Dame Evelyn Glennie said, "Silence is a sound." And in examining its role, we can start to appreciate how vital a sound it is.

Just as mountains need valleys to define them, all sounds need silence to give them context, structure and meaning. It's the gaps between the words and the notes that create language and music. Without silence, there is only meaningless cacophony. With silence, there can be heightened consciousness: the Japanese tea ceremony is performed in silence so that the participants can savor the tiny noises of stirring and pouring, which would normally be drowned by conversation.

It may seem counterintuitive to celebrate silence in a book where I am arguing for us all to awaken to the importance of sound and savor it consciously, but I must, for a variety of reasons.

Silence creates rhythm, one of the most fundamental aspects of sound for all species, because rhythm implies life, agency and possibly intention. It does this both by defining the pattern of sounds and by anchoring one end of the continuum of intensity. A skilled actor or orator is best friends with silence, using it for emphasis, shock, variation and to amplify meaning. One of the commonest failings I see in inexperienced and especially very nervous public speakers is their fear of silence, which drives them to fill every available moment with gabble or meaningless filler words. A top salesperson is as assured and adroit with silence as an expert surgeon with a scalpel, especially when it comes to the close. The same goes for elite negotiators. Silence is one of the main reasons speaking is so much more powerful than writing: it's very hard to write silences. As anyone who has ever given a speech or presentation knows, the silence of impressed awe and disengaged boredom may be the same sound, but they feel very different. Silence is not homogeneous.

People often see silence as an absence of fun, hence restaurant and bar proprietors confusing noise with enjoyment. The circumplex psychological model of human emotions, widely used in research into the effects of music on people, stresses that arousal, like chocolate cake, has diminishing returns: in excess, both make people unhappy and ill. The competing voices of entertainment, advertising, social media and the internet, delivered through always-on devices, are edging our whole society ever further down a road to an addiction to intensity and ultimately psychosis. We need to start listening before we go deaf—and silence is the perfect antidote.

I spoke to George Prochnik, author of the excellent book *In Pursuit of Silence*, at his home in Brooklyn, and he talked of his sense that our world is becoming more and more frenetic and noisy. He had recently visited a friend whose children were watching *Sesame Street* and had been struck by how intense the sound was. He said:

> It was so loud and so jangly and so unbelievably stimulating that I went back and I rented the earliest iteration of this program that had ever been made, and by comparison it was like *Sesame Street* on Quaaludes. It was so gentle: the children were playing in an absolutely engaged way... It wasn't so souped-up and it wasn't so unbelievably over-stimulating. One of the silent markers, I think, for the oversaturation of noise in our society today may be this epidemic of attention deficit disorders and hyperactive disorders and all of that band of problems. I'm certain that the level of sonic stimulation (and visual stimulation as well) is part of the reason that this is happening. If you put those two programs next to each other, you just felt: how could a child not become hyper in response to the more recent version?

However, in the same way that there's that incredible sensitivity to an abundance of acoustical stimulation, people have been finding that very quickly you can start to bring some sensitivity to sound with children in a way that can linger and affect their attitudes. For example, there's a woman I spoke with who works with disadvantaged youth from often very noisy homes and noisy neighborhoods—children who have very little knowledge of what silence is. And she began taking them on silent walks through the countryside. She'd take them out of the city for the day and demand that they did these hikes in silence. And at first there's a lot of giggling, and difficulty in sustaining that quiet, but she said that even kids who've virtually never known silence in their homes, by the end of this experience were not only very positively engaging with the small voices that make up any natural environment but were excited about the prospect of doing something like this again. The associations that we have culturally with silence are often so pious, at the very least solemn. And I think we really need to expand our range of definitions for what silence enables in order to make it something that could have an appeal to young people.

I agree with George, and if we fail to persuade more people of the value of silence, there is a danger that tranquility will become the exclusive preserve of the rich. A recent marketing trend has been the replacement of time with peace: rather than promoting products that save time and allow you to pack more into your day, advertising campaigns have appeared with slogans like "the quietest cabin in the sky." Silence is a universal right, not a luxury item, though it is increasingly rare.

Discovering silence

So where is silence to be found in the modern world? And once found, is it possible to defend it? One man who made it his life's work to push back against intrusive noise is Gordon Hempton. His fight to protect the pristine soundscape of the Hoh Rainforest in Olympic National Park, Washington State, gave rise to the book *One Square Inch of Silence: One Man's Quest to Preserve Quiet* and the documentary film *Soundtracker*. The one square inch Hempton started defending in 2005 is marked by a small red stone, and the fight to protect its soundscape has a wide impact. Silence spreads, just like noise: if Hempton and his supporters could assure that one square inch of silence, it would have implications for perhaps twenty kilometers in all directions, which is a surface area of 1,257 square kilometers.

Absolute silence is, of course, not really Hempton's goal. In line with our own looser definition of the word, in a podcast interview with *On Being*'s Krista Tippett he clarified: "When I speak of silence, I often use it synonymously with quiet. I mean silence from modern life, silence from all these sounds that have nothing to do with the natural acoustic system, which is busy communicating; wildlife are as busy communicating as we are. But it's not just messages coming from wildlife—and I can name some that have been really transformative in my personal life—but it's also the experience of place, what it means to be in a place."

Hempton secured cooperation from the park authorities to protect his one square inch, and for years it was unsullied, apart from the odd flight out of SeaTac Airport. Then, in 2018, the U.S. Navy spotted that population density was very low in the park and designated it a low-flying exercise area on the basis that fewer people would be upset by the noise. Today, they fly thousands of fast jet sorties over the park every year and the silence is, in Hempton's words, "practically extinct."

When I spoke with Gordon at his home in Seattle, I had to ask him: what do you do when your life's dream is destroyed like that? He replied:

> I went to the quiet. One of the biggest benefits of quiet experience is that you can ask impossible questions of the quiet and you will receive an answer. And the question I asked the quiet was: "Where did I go wrong?" And to my surprise, the quiet laughed at me... I could feel it come up in my own chuckle, right in my chest. Then I began to laugh. And it was like, "Come on, really? It's obvious!" Why should the world care about a place that they may never visit when quiet is a birthright, when quiet is something that is essential to being a human being, to thinking your own thoughts and feeling your own emotions, which essentially defines who you are. Those moments of quiet are the think tank of the soul, and in that one moment in the Hoh Valley, Quiet Parks International was born.

Washington State's loss became the world's gain. Quiet Parks International [113] is a charity that aims to create or encourage the formation of quiet places all over the world. It does this directly by setting them up, and indirectly by setting international standards and giving "QPI Awards" to places that meet them, be they wildernesses, urban parks or trails. The website has a map showing quiet places nominated by likeminded people all over the world, including places to stay (the ones that QPI has assessed and awarded or certified are differentiated). The co-founder of QPI, Tim Gallati, has also established an invaluable online library of over twenty-five thousand research articles on noise and quiet, publicly accessible [114]. With all this activity, it's understandable that Gordon Hempton has faith in the future. He told me, "As something begins to disappear, it becomes more valuable and then we design our future. Anybody who's not optimistic about the future, all I can say is—go find quiet." This sounds enticingly simple, but I suspect our increasingly populous, urbanized, tech-obsessed human world will make finding quietness more and more challenging for most people—which of course makes initiatives like QPI all the more valuable.

The most important silence

Another seeker of quietness who has had a major global impact is Susan Cain. I met Susan just before she gave her famous TED talk (based on her book *Quiet*) about the importance of valuing introverts, those people who prefer quietness and solitude, and often get overlooked in any society that confuses "sound and fury" with significance. The talk went viral, and the combination of the book and TED created a movement that changed many people's lives. Susan told me, "Over the past decade hundreds, perhaps thousands, of companies and schools have made changes to their hiring, promotion, diversity and inclusion, and educational practices, to harness the talents of introverted students, employees, leaders, and creatives." Making time to listen to quiet people is an essential discipline for any effective leader, as is creating the spaces that will allow them to be most productive.

Of course, listening is much easier to do in a quiet space—but the most important silence required for effective listening is that of the listener. In human communication, the most common obstacle to this critical silence is a habit I call speechwriting: while someone is talking, the supposed listener is in fact composing their next brilliant contribution to the conversation, listening far more attentively to the voice in their head than to the one in front of them. Since we can think much faster than we can speak, this habit often results in impatient interrupting, another classic destroyer of relationships.

The silence of a true listener is close to meditation. It takes some courage, because it requires faith that, when it becomes one's turn to speak, the right words will arrive without being prepared, edited or practiced. It also takes (and develops) humility, because giving the gift of one's total attention is inherently selfless, generous and an acknowledgment of the other person's at least equal importance. Finally, it requires setting aside the need to be right, so that bias, denial or opinion do not block challenging but possibly useful communication. For this reason, it often becomes difficult for leaders of organizations and communities, or fêted individuals who are used to adulation, to listen well: the ego is never a good listener. Sadly, we see this increasingly in global politics. Our leaders often meet for talks. It might just be a better and safer world if they met for listens instead.

True listening is like the quiet contemplation that's at the heart of every spiritual path of which I am aware. The founders and leaders of every religion have always retired to solitary, silent places of listening to receive the wisdom they later share. Effective prayer is rooted in listening.

However, listening is under pressure in this overstimulated, fast-paced world. Most people's response to the noise around them is to try to speak more effectively—or just be louder. As noted, my TED talk on speaking has five times as many views as my one on listening, which is a fair barometer of the weight we attach to each of these skills.

I hope that in joining me on this journey through the wonders—and the horrors—of sound, you have found new reasons to listen more carefully and consciously. There are nine billion of us crowded onto this planet now. We are now so numerous that the sum of our individual decisions impacts massively on both the climate and the ecosystem. We have invented multiple methods of potential mass destruction—from weapons and pathogens to AI—and they are not all in safe hands. Those of us fortunate enough to live free from war, famine, poverty, disease or repressive regimes tend to live in a complacent bubble, assuming that tomorrow will be similar to today. Some of the loudest sounds we've encountered in this book show us that that assumption is not always a reliable one.

Our survival, individually and as a species, has always depended on how well we listen. Nothing has changed. I strongly believe that never has listening been more important—or less valued—than it is today. I hope that this book has unlocked a new appreciation of sound in you, and that you will be listening afresh, with renewed wonder and respect, to the world, its living things, and especially other people.

Re-engaging with silence

Living in a city can create a numbness of hearing, and especially listening, because of the constant noise, as well as a disconnection from nature because natural sounds are rare unless it rains or there is a thunderstorm. Though some cities offer oases of tranquility, these usually have to be sought out and often need to be consciously protected. Each city has its own voice: I remember my wild excitement when I first visited New York aged sixteen and heard the wailing police sirens in the street below. At that time the UK's police cars used two-tone horns, so this was hugely stirring to me. I excitedly exclaimed: "I'm in *Kojak*!" You may love your own city's reassuring sonic signature but, still, I suggest giving yourself breaks.

WHAT YOU CAN DO

A practice of sitting in silence for a few minutes, ideally several times a day, is a wonderful way of improving your listening, as well as your general well-being. Recording engineers must take breaks like this, because after a few hours of listening, especially at high volume, the whole sense of hearing dulls, and it becomes impossible to discern what sounds good; the result is usually an endless series of turning things up until the mix is a complete mess. Silence is the context and the baseline for sound, so reacquainting your ears with it like this refreshes and recalibrates them, allowing you to listen with renewed accuracy and clarity.

This is especially important for anyone who has lost the connection with silence to the point where it makes them uncomfortable. If that's you, try a very gentle introduction, starting with just one minute then building up by thirty seconds a day. Of course, the quiet that's achievable for most people in cities is limited: it'll probably be found in a bedroom in a home, or a meeting room in an office. If you can get away from concrete and into nature, the options are far greater.

I strongly recommend such a practice of silence. To help with your first try, here are the lovely meditations one encounters on small signs around the Way of Silence, the path that runs around the convent and twelfth-century Basilica on the island of San Giulio in Lake Orta, in Piedmont, northwestern Italy. The nuns in the convent live in complete silence, and the signs suggest that one undertakes the walk around the Way of Silence in the same state. The meditations from Abbess Mother Anna Maria Canopi are surprisingly eclectic. I suggest you spend at least thirty seconds on each one to start with.

> In the silence you accept and understand
> Silence is the language of love
> Silence is the peace of oneself
> Silence is music and harmony
> Silence is truth and prayer
> In the silence you meet the Master
> In the silence you breathe God
> Walls are in the mind
> The moment is present, here and now
> Leave yourself and what is yours
> In the silence you receive all

Savoring silence

There are distinct varieties of silence. I still clearly recall three very different silences that I experienced, many years after each event. The first occurred one warm afternoon sitting on a scree slope high on a Dolomite mountain called Civetta (Owl). I was far above the tree line, so there were no birds, and the afternoon was windless. The silence went on forever, like the view, endless and vast. Another type of silence altogether was the stifling one that pressed in on me deep in Torquay's Kents Cavern when we switched off the lights. And a third was the serene, spacious silence of the extraordinary Worth Abbey church in West Sussex. This huge monastic church, designed by Francis Pollen, resembles an upturned bowl; its circular stone interior seats nine hundred people, but at night it is empty and still, with a single beam of light illuminating the altar in the center. The silence there is deep, calming and comforting.

I hope you have your own cherished memories of tranquil places. Visit them often, either physically or mentally, and establish your own practice of silence to reconnect with the still center of all things. Life is richer and more balanced when silence is part of it.

Fortunately, it's not just a few visionaries who are evangelizing silence, or at least tranquility. There is a growing global movement that encourages spending attentive time in quiet nature. Of course, this is hardly a new idea—probably since the dawn of our species, people have enjoyed quiet walks or sitting and listening to wind in leaves and birdsong—but in Japan, where the movement started, it was rebranded and rejuvenated in 1982 as *shinrin-yoku* (森林浴) which means forest bathing. This simply involves walking calmly and quietly in a forest, breathing deeply, and consciously experiencing nature in all five senses.

Shinrin-yoku centers are proliferating not only in Japan, where they build on centuries of religious and philosophical appreciation of the importance of forests, but all over the world. There are books, websites, guides and trainers in many countries, and guided walks are on offer in places like Adirondack Park, upstate New York, the rainforests of Costa Rica, New Zealand's Waipoua Forest, Kenya's Great Rift Valley and the lush forests of Hawaii. Some even offer virtual forest bathing, though it is unclear how beneficial it can be to watch someone else walking through nature—but perhaps something is better than nothing at all for people who simply can't reach beautiful nature. In the U.S., the Association of Nature and Forest Therapy [115] was founded in 2012, while in the UK, The Forest Bathing Institute [116] works with universities to generate new research on the practice. There are millions of people worldwide exploring this antidote to dense, noisy urban life. But does it work?

The research is broadly positive. Forest bathing does seem to lower physiological stress markers (cortisol, pulse rate and blood pressure) and clearly many people enjoy it, feeling that it enhances their well-being and happiness. Mostly it's recommended as preventative therapy, though there is little as yet assessing its effectiveness in moderating or curing any specific conditions. But anecdotes like George Prochnik's about the kids on silent walks are common, and it makes sense intuitively that people with a range of mental health issues who might find mindfulness, meditation or other inward-focused practices very challenging may find focusing their attention *outwards* on beautiful, calming natural environments enormously soothing and beneficial.

WHAT YOU CAN DO

Whether you explore formal forest bathing or simply take a quiet walk in whatever nature is available to you, I am certain that you will feel the benefit. Trees are not the only inspiring natural objects: meadows, streams, rivers, lakes, mountains and seashores could all be given the same marketing makeover that forests have received, and walking in those places could yield many of the same benefits for physical and mental health, as well as pure enjoyment—and, most important of all, a reconnection with nature, a reforging of respect for other living things and for our planet, and a renewed appreciation of the importance of tranquility in an increasingly noisy, over-stimulated world. The world is sound, and silence makes sense of sound. I urge you to embrace it, adopt your own practice of silence, and teach your children the importance of quietness in a balanced life.

*

Each morning, we open our eyes and the darkness of sleep gives way to vision. We can all benefit from recreating this dynamic in our relationship with sound, and then holding on to that attentiveness. To do this, we must practice it throughout our day in meetings, conversations and calls with others as we move around, as we encounter the sounds of humanity, of other creatures or of our planet.

Identify the sounds that make you happy, support your well-being or help you to think or work—and those that challenge you. Start to replace the harmful sounds with helpful ones. Most of all, give the precious gift of true listening to others—listening consciously, committedly, with compassion and curiosity, seeking to understand, even if we disagree (often especially if we disagree). This is transformative.

Sound affects you, and you affect sound. Conscious listening is the only way to understand and fully experience this critical relationship, and thus to take responsibility for the sound you create, and the sound you consume.

Put this book down. Close your eyes. Listen.

APPENDIX

Table of disgusting sounds

Rank	Sound
1	Vomiting
2	Microphone feedback
3=	Multiple babies
3=	Scrape/squeak (like train wheels)
5	Squeak (like a seesaw)
6	Violin
7=	Whoopee cushion
7=	Baby cry
9=	Soap opera argument
9=	Mains hum
11	Tasmanian devil
12=	Cough
12=	Cat spitting and howling
12=	Mobile phone rings
15	Creaky door
16=	Barking mad dog
16=	Sniff
16=	Fingernails scraping down a blackboard
16=	Polystyrene
20	Dentist's drill
21	Cough & spit
22	Alarm clock
23	Fast electrical drilling
24	Apple munch
25	Creaky door
26=	Squeaky trolley
26=	Snoring
28=	Electrical throb
28=	Cat eating noisily
30	Whoopee cushion reverberated
31	Aircraft take-off
32	Drums
33	Gong
34	Low not-quite-eerie noise

Trevor Cox, "Scraping sounds and disgusting noises," *Applied Acoustics*, Volume 69, Issue 12, 2008, pp 1195–1204

SCIENTIFIC REFERENCES

For those of you with a scientific leaning, there is a list of the scientific papers and books that underpin the facts and theories discussed throughout the book on the book's website at www.soundaffectsbook.com.

FURTHER INVESTIGATIONS

If you want to continue your journey into sound by going deeper into any of the topics covered in this book, I have compiled a list of books, films and music that you might like to explore. This list also exists on the book's website with links to save you the detective work.

1 Giant Leap, Jamie Catto and Duncan Bridgeman, Palm Pictures, 2002

A Book of Noises: Notes on the Auraculous, Caspar Henderson, Granta, 2023

A Book of Silence: A Journey in Search of the Pleasures and Powers of Silence, Sara Maitland, Granta, 2008

Acoustic Communication, Barry Truax, Praeger Publishers Inc, 2000

Alone Together: Why We Expect More from Technology and Less from Each Other, Sherry Turkle, Basic Books, 2017

Archaeoacoustics: The Archaeology of Sound, ed., Linda C. Eneix, OTS Foundation, 2014

Auditory Neuroscience: Making Sense of Sound, eds., Jan Schnupp, Israel Nelken and Andrew King, MIT Press, 2012

Auditory Scene Analysis: Perceptual Organization of Sound, Albert S. Bregman, MIT Press, 1994

Bug Music: How Insects Gave Us Rhythm and Noise, David Rothenberg, St. Martin's Press, 2013

Case Studies in Music Therapy, ed., Kenneth E. Bruscia, Barcelona Publishers, 1991

Deep Listening: A Composer's Sound Practice, Pauline Oliveros, iUniverse, 2005

Designology: How to Find Your PlaceType & Align Your Life with Design, Sally Augustin, Mango, 2019

Discord: The Story of Noise, Mike Goldsmith, Oxford University Press, 2014

Handbook of Music and Emotion: Theory, Research, Applications, eds., Patrik N. Juslin and John A. Sloboda, Oxford University Press, 2011

Handbook of Music Psychology, ed., Donald A. Hodges, IMR Press, 1996

Haunted Weather: Music Silence and Memory, David Toop, Serpent's Tail, 2005

Hearing Solar Winds, David Hykes and the Harmonic Choir, Ocora, 1983

How to Be Heard: Secrets for Powerful Speaking and Listening, Julian Treasure, Mango, 2017

How to Speak So That People Want to Listen, Julian Treasure, TED talk, 2014

How to Talk with Anyone About Anything: The Practice of Safe Conversations, Harville Hendrix and Helen LaKelly Hunt, Thomas Nelson, 2024

In Pursuit of Silence, dir., Patrick Shen, Transcendental Media, 2016

In Pursuit of Silence: Listening for Meaning in a World of Noise, George Prochnik, Anchor Books, 2010

Into a Wild Sanctuary: A Life in Music and Natural Sound, Bernie Krause, Heyday Books, 1997

Just Listen: Discover the Secret to Getting Through to Absolutely Anyone, Mark Goulston, AMACOM, 2009

Krakatoa: The Day the World Exploded: 27th August 1883, Simon Winchester, Penguin, 2003

Listening (Perspectives in Continental Philosophy), Jean-Luc Nancy, Fordham University Press, 2009

Listening for Ancient Gods: Archaeoacoustics: A study of the world's oldest buildings and the archeology of sound, with new implications for how we came to be who we are, Linda C. Eneix, CreateSpace Independent Publishing Platform, 2016

Listening for Democracy: Recognition, Representation, Reconciliation, Andrew Dobson, Oxford University Press, 2014

Listening in Everyday Life: A Personal and Professional Approach, eds., Michael Purdy and Deborah Borisoff, University Press of America, 1996

Music from the Galaxies, Fiorella Terenzi, Universal/Mercury/Polygram, 2009

Musical Illusions and Phantom Words: How Music and Speech Unlock Mysteries of the Brain, Diana Deutsch, Oxford University Press, 2019

Musicophilia: Tales of Music and the Brain, Oliver Sacks, Alfred A. Knopf, 2008

Mysterium Tremendum, Mickey Hart Band, 360° Productions, 2012

Nada Brahma—The World Is Sound: Music and the Landscape of Consciousness, Joachim-Ernst Berendt, Destiny Books, 1988

Noise, Bart Kosko, Viking, 2006

Noise and Its Effects, eds., Linda Luxon and Deepak Prasher, John Wiley & Sons, 2007

Noise: A Human History of Sound and Listening, David Hendy, Profile Books, 2013

Noise and Man, William Burns, John Murray, 1968

On Listening, eds., Angus Carlyle and Cathy Lane, Uniformbooks, 2011

One Square Inch of Silence: One Man's Search for Natural Silence in a Noisy World, Gordon Hempton, Simon & Schuster, 2009

Quantum Listening, Pauline Oliveros, Spiral House, 2022

Quiet: The Power of Introverts in a World That Can't Stop Talking, Susan Cain, Crown Publishers, 2012

Reclaiming Conversation: The Power of Talk in a Digital Age, Sherry Turkle, Penguin Press, 2015

Rhythms of the Brain, György Buzsáki, Oxford University Press, 2006

Seeing Silence: The Beauty of the World's Most Quiet Places, Pete McBride, Rizzoli, 2021

Sensehacking: How to Use the Power of Your Senses for Happier, Healthier Living, Charles Spence, Viking, 2021

Sensory Evaluation of Sound, ed., Nick Zakharov, CRC Press, 2020

Silence: Lectures and Writings, John Cage, Marion Boyars, 1968

Songs of the Humpback Whale, Roger Payne, Living Music, 1991

Sonic Alchemy: Conversations with Leading Sound Practitioners, Joshua Leeds, InnerSong Press, 1997

Sonic Experience: A Guide to Everyday Sounds, eds., Jean-François Augoyard and Henry Torgue, McGill-Queen's University Press, 2005

Sonic Meditations, Pauline Oliveros, PoP and MoM Publications, 2022

Sonic Wonderland: A Scientific Odyssey of Sound, Trevor Cox, Bodley Head, 2014

Sound Business: How to Use Sound to Grow Profits and Brand Value, Julian Treasure, Management Books, 2011

Sound Design: The Expressive Power of Music, Voice and Sound Effects in Cinema, David Sonnenschein, Michael Wiese Productions, 2001

Sound Ideas: Music, Machines and Experience, Aden Evans, University of Minnesota Press, 2005

Sounds Wild and Broken: Sonic Marvels, Evolution's Creativity and the Crisis of Sensory Extinction, David George Haskell, Faber, 2022

Soundtracker, dir., Nicholas Sherman, Fou Films, 2010

Spaces Speak, Are You Listening? Experiencing Aural Architecture, Barry Blesser and Linda-Ruth Salter, MIT Press, 2009

Sun Rings, Terry Riley/Kronos Quartet, Nonesuch, 2019

Survival of the Beautiful: Art, Science and Evolution, David Rothenberg, Bloomsbury, 2011

Sweet Anticipation: Music and the Psychology of Expectation, David Huron, MIT Press, 2006

Sync: The Emerging Science of Spontaneous Order, Steven Strogatz, Penguin, 2004

The Book of Music and Nature: An Anthology of Sounds, Words, Thoughts (Music / Culture), eds., David Rothenberg and Martha Ulvaeus, Wesleyan University Press, 2001

The Great Animal Orchestra/Carnival of the Animals, Richard Blackford, Nimbus Records, 2015

The Great Animal Orchestra: Finding the Origins of Music in the World's Wild Places, Bernie Krause, Little Brown & Co., 2012

The Handbook of Hearing and the Effects of Noise: Physiology, Psychology and Public Health, ed., Karl D. Kryter, Elsevier Science Publishing, 1994

The Listening Book: Discovering Your Own Music, W. A. Mathieu, Shambhala Publications, 1991

The Lost Art of Listening: How Learning to Listen Can Improve Relationships, Michael P. Nichols, Guildford Press, 2009

The Music Instinct: How Music Works and Why We Can't Do Without It, Philip Ball, Bodley Head, 2010

The Musical Mind: The Cognitive Psychology of Music, John A. Sloboda, Oxford University Press, 1985

The Mysticism of Sound and Music: The Sufi Teaching of Hazrat Inayat Khan, Hazrat Inayat Khan, Shambhala Publications, 1996

The Power of Sound: How to Be Healthy and Productive Using Music and Sound, Joshua Leeds, Healing Arts Press, 2001

The Power of Tranquility in a Very Noisy World, Bernie Krause, Little Brown and Co., 2021

The Sacred Art of Listening: Forty Reflections for Cultivating a Spiritual Practice, Kay Lindahl, SkyLight Paths Publishing, 2002

The Singing Neanderthals: The Origins of Music, Language, Mind and Body, Steven Mithen, Weidenfeld & Nicolson, 2005

The Social and Applied Psychology of Music, Adrian North and David Hargreaves, Oxford University Press, 2008

The Songs of Trees: Stories from Nature's Great Connectors, David George Haskell, Viking, 2017

The Sound Studies Reader, ed., Jonathan Sterne, Routledge, 2012
The Sounds of Life: How Digital Technology is Bringing Us Closer to the Worlds of Animals and Plants, Karen Bakker, Princeton University Press, 2022
The Soundscape of Modernity: Architectural Acoustics and the Culture of Listening in America, 1900–1933, Emily Thompson, MIT Press, 2004
The Soundscape: Our Sonic Environment and the Tuning of the World, R. Murray Schafer, Destiny Books, 1994
The Space Project, Various Artists, Lefse Records, 2014
The Tao of Music: Sound Psychology, John M. Ortiz, Weiser Books, 1997
The Third Ear: On Listening to the World, Joachim-Ernst Berendt, HarperCollins, 1988
The Universal Sense: How Hearing Shapes the Mind, Seth Horowitz, Bloomsbury, 2012
The Wisdom of Listening, ed., Mark Brady, Wisdom Publications, 2003
The World of Silence, Max Picard, Gateway Editions, 1972
Thinking in Sound: The Cognitive Psychology of Human Audition, eds., Stephen McAdams and Emmanuel Bigand, Science Publications, 1993
This Is Your Brain on Music: Understanding a Human Obsession, Daniel Levitin, Atlantic, 2007
Thousand Mile Song: Whale Music in a Sea of Sound, David Rothenberg, Basic Books, 2008
Touch the Sound: A Sound Journey with Evelyn Glennie, dir., Thomas Riedelsheimer, Filmquadrat/Skyline Productions, 2004 (Signum DVD)

Water Sound Images, Alexander Lauterwasser, Macromedia Publishing, 2011

Ways of Hearing, Damon Krukowski, MIT Press, 2019

Whale Music: Thousand Mile Songs in a Sea of Sound, David Rothenberg, Terra Nova Press, 2023

What About Me?, 1 Giant Leap, Channel 4, 2008

Why Birds Sing: A Journey into the Mystery of Birdsong, David Rothenberg, Basic Books, 2005

You Are What You Hear: How Music and Territory Make Us Who We Are, Harry Witchel, Algora Publishing, 2010

Zero Decibels: The Quest for Absolute Silence, George Michelsen Foy, Scribner, 2010

ACKNOWLEDGMENTS

First, to Jane, Holly and Sapphire for their love and support throughout the birthing of this book, which from the beginning felt like the most significant thing I will do in my life and took me away from them for countless hours.

To my wise and wonderful mentor and agent, Richard Pine at Inkwell, who made everything happen in the first place and put bread on our table throughout the years it took to go from concept to publication.

To the brilliant teams at Quercus and Grand Central, with special and profound gratitude to my editors, Richard Milner in London and Karyn Marcus in New York, for their patience, encouragement and excellent judgment, and to Jamie Coleman and Ian Preece for their invaluable inputs and enthusiasm; also to Quercus MD Jon Butler for all his support—and not to forget the original editor, Katie Follain, for starting the ball rolling in the first place.

To my dear friend and sound pioneer, Bernie Krause, for his invaluable comments and advice as the text developed.

And finally to all the sound lovers and listeners I have encountered—those I have met, interviewed, conversed with or made music with; those who have inspired me with their passion and genius; and those who have taught me the incomparable value of listening in living a conscious life.

INDEX

Page numbers in *italics* indicate illustrations and tables

Absolute Radiometer for Cosmology, Astrophysics, and Diffuse Emission (ARCADE), 257, 262–3
acoustics, 45–6, 175–85
    acoustic environments, 4–5, 69
    acoustic monitoring, 173–4
    architectural, 8, 175–84, 213–14
active sonar 156
ADSR (attack, decay, sustain, release), 64
Advanced Laser Interferometer Gravitational-Wave Observatory (LIGO), 281
Aeolian tones, 244
aeroacoustics, 244
AI (Artificial Intelligence), 101–2, 173–4, 195, 203, 210, 219
air-conditioning, 46–7
aircraft noise, 160–62
Alaska, 141, 253
Almagest (Ptolemy), 295
Alone Together (Turkle), 7
Amazon, 197, 206, 209
American Academy of Nursing, 164–5
Amish community, 155
amplitude, 57, *57*

analogue recording, 205
Anderson, Marc, 115
André, Michel, 173–4
anechoic chambers, 3–4, 301
Angel Falls, Venezuela, 249
animal sounds, 73–5, 109–49
    apes and bears, 122–3
    cave-dwelling species, 110
    echolocation, 111–12, 136–7
    head size and, 109–10, *110*
    hearing range, 125, *125*
    infrasound, 115–17
    marine creatures, 126–39
    pets, 118–21
    reptiles, 117–18
    ultrasound, 109, 113
Antarctic, 111, 161, 243, 248
anthropophony, 151–81, 282
    animals and, 169
    see also archaeoacoustics; noise
ants, 89
apes, 122–3
Apple, 197, 206, 209
aquatic mammals, 128–30, 132–3
arachnids, 91
archaeoacoustics, 9, 175–9, 193

Archer, Martin, 274–6
Arctic, 123, 156, 248
Areni, Charles, 41
Aristotle, 49, 266
Arup (design and engineering firm), 184
asteroseismology, 268
astrology, 295
astronomy, 54–5, 68, 147–8, 294–8
    aural astronomy, 270–74
    see also cosmos/cosmophony
Atkinson, Max, 124
atomic force microscope (AFM), 76–7
Attenborough, David, 89, 105
attentive listening, 10–11
audio technology, 6–7, 173–4, 205–15, 218–20
"auditory learning," 100
Auditory Neuroscience (Schnupp, ed.), 19
Augoyard, Jean-François, 69
Automatic Dialogue Replacement (ADR), 201
Ayres, José Márcio, 173

bacteria, 79–80
Baghdasaryan, Davit, 210
Baker, Ed, 93
Bakker, Karen, 101, 149
Banbury, Simon, 37
Baptista, Luis, 107
Barkhausen, Heinrich, 292
Barrow Island, Australia, 242–3
Bartel, Lee, 80
bats, 88, 111–12
BBC (British Broadcasting Corporation), 92, 94
beam steering, 213–15
bears, 123
Beautiful Now website, 114–15
Beaver, Paul, 140
bees, 82–3, 90, 92–3
Beethoven, Ludwig van, 194
Bell, Alexander Graham, 58

Benway, Evan, 221
Beranek, Leo, 181
Berendt, Joachim-Ernst, 10
Berger, Jonathan, 198
Berry, Dianne, 37
Beston, Henry, 225
Bhosle, Asha, 191
Big Bang theory, 258–9, 261
BioAcoustica (online sound resource), 93
bioacoustics, 141
biomass, 81–2, 82
biophilia, 36, 44–5, 48, 220–23
biophony, 71, 73–5, 106, 140–45, 282
    environment and, 106
    see also animal sounds; birds/birdsong; cells; insects; plants
birds, 94–108, 117, 169
    birdsong, 35–6, 94–100, 103–4
    environment and, 106
    harmony, 99
    hearing, 97
    mimicry, 104–6
    regional accents, 106–7
    rhythm, 98–9
    syrinx, 98
black holes, 12–13
Blackford, Richard, 142
Blackwell, Chris, 191
Blesser, Barry, 181–3
Blitz, Leo, 54–5
Boeri, Stefano, 221
Boethius, 68
bone conduction, 207–8
bouba/kiki effect, 44
Bourdeau, Ethan, 183
Brahe, Tycho, 294–5
brain, 20–24
    amygdala, 20
    cortex, 20
    musicians and, 24
    reticular formation, 23
    sonic patterns and, 22–4

Brandenburg, Karlheinz, 197
Bridgeman, Duncan, 191
Brumm, Henrik, 106
Bug Music (Rothenberg), 149
Busoni, Ferruccio, 189
buzz pollination, 82–3
Byrne, David, 217

Caesar, Julius, Roman Emperor, 152
Cain, Susan, 8, 307
Californian singing fish, 131
Call of Duty: Warzone (game), 207
Cameron, James, 207
cancer, 77–9
Canopi, Anna Maria, 311
Carancas, Peru, 237
Carter, Brandon, 148
Cassini space probe, 286–7, 289, 292
cats, 113, 118–19
Catto, Jamie, 191–2
Cave, Nick, 219
cave swiftlets, 102
cells, 76–80, 144
Cetacean Translation Initiative (CETI), 101
cetaceans, 129, 133, 156–7
Chandler, Simon, 34
Chandra space telescope, 13, 271–2, 279
Chanson, Hubert, 252
Chicago Symphony Orchestra, 30
Chicxulub crater, Mexico, 94, 232, 237, 254
children
    babies, 17–18
    education, 161
    silence and, 304
chimpanzees, 101, 122
Churchill, Winston, 185
cicadas, 86–8, 90, 124, 132
cities, 8, 153–4, 162–3, 167–9, 221, 310
Clark, Christopher, 159
cockroaches, 89
"cocktail party effect," 32

Coimbra, Fernando, 177
communication, 6–7, 9
Condé Nast Traveller, 162
conversation, 7, 38–9, 87, 143, 221
Cook, Ian, 178
Copernicus, Nicolaus, 294
Cornell University, 94, 95, 105
Cosmic Microwave Background (CMB), 261–2
cosmos/cosmophony, 257–99
    black holes, 279, 280–81
    electromagnetic radiation, 292–3
    extraterrestrial sound, 282–92
    galaxies, 265
    gravitational waves, 279–82
    gravity, 269–70, 280–81, 297
    harmony of the spheres, 294–9
    magnetic fields, 274–6, 278
    music and, 272–3
    plasma, 259–62, 266–7, 269–70, 275
    pulsars, 278
    solar system, 276, 282–91, 296–8
    solar winds, 274–5
    "sonification," 271–4
    soundwaves and, 259–60, 264–5, 269
    "space roar," 258–65
    stars, 265, 267–9, 277–8
    supernovae, 276–8
Cousteau, Jean-Michel, 159
Covid-19 pandemic, 162–3
Cox, Trevor, 178
coyotes, 122
crickets, 86–9, 96, 115
crocodiles, 117–18
crustaceans, 127, 130–32
Csíkszentmihályi, Mihály, 38
Cue, Eddy, 206–7
Curzon, George, 1st Marquess Curzon of Kedleston, 236

Darwin, Charles, 74, 96
De Rijke, Victoria, 106
deafness, 29

Debertolis, Paolo, 177–8
decibels, 57–61, 132n
deer, 119–20
Delhi, India, 166
Descent of Man, The (Darwin), 96
Descript app, 210–11
Dhaka, Bangladesh, 167
Díaz-Merced, Wanda, 273
diegetic sound, 201
digital signal processing, 218
digital sound, 196–9, 205, 208
directional sound, 215
Discord: The Story of Noise (Goldsmith), 163
disintermediation, 220
dogs, 120–21
Dolby Atmos, 200, 206–7
dolphins, 88, 99, 105, 129–30, 133–7, 157
Dominici, Francesca, 160
Dry Falls, Washington State, 251
Dubai Fountain, Burj Khalifa Lake, 223
ducks, 106
Dugu, Portia Domonatani, 231
"Dusk by the Frog Pond" (Anderson), 115
Dwarfie Stane, Hoy, 176–7

Earle, Sylvia, 159
ears, 16–19, 25, 29, 127
    in-ear technology, 209
Earth, 227–9, 296–7
    birth of, 227–9
    Earth hum, 244–5, 282–3
    geology of, 229–30, 232
    see also geophony
Earth Impact Database, 239
earthquakes, 36, 233–5, 253
echolocation, 102, 136–7, 111–12
Edison, Thomas, 195
8oUA (Fabrica Music Area), 271
Einstein, Albert, 298
elephants, 115–16
Enceladus (Saturn's moon), 287–8

Eneix, Linda, 177, 179
Eno, Brian, 218
entrainment, 28–9, 40
envelope, 64–5
environments, 45–9, 220–22
    acoustics, 45–6, 48–9
    masking sounds, 46–7
    multi-sensory, 44
    noise level, 48
    running water and, 222–3
    silence and, 312–13
    workplace, 221–2
eukaryotes, 75
Europa (Jupiter's moon), 288
European Space Agency (ESA), 271, 277
European Union Environmental Noise Guidelines, 165
Event Horizon telescope, 279
Everything Is Amazing (Sowden), 188
evolution, 146–8
extraterrestrial life, 147–8, 282

Fabrica Music Area, 271
Farbod Alijani, 79
Fazenda, Bruno, 178
film industry, 200–203
Final Fantasy (game), 203
fish, 127–8, 130–31
Fishlake National Forest, Utah, 242
Fleming, Kate, 124
flies, 90, 92
Foley, Jack, 201
forest bathing, 312–14
forest fires, 241–2
Forman, William, 279
Freeman, Angie, 117
frogs, 96, 113–15, 124
Frontiers 2022: Noise, Blazes and Mismatches—Emerging Issues of Environmental Concern, 167
Fuller, Jim, 268–9

Galaxie Cygnus-A (Schröder), 272
Galileo space probe, 288, 292
Gallati, Tim, 307
gaming, 203, 207
Ganymede (Jupiter's moon), 288
Gardner, Peter, 78
geese, 98
Geissenklösterle cave, Germany, 193
geophony, 225–53
    Earth's hum, 244–5
    earthquakes, 36, 233–5, 253
    forest fires, 241–2
    lightning, 241
    meteor/asteroid impacts, 232, 236–9
    sand dunes, 235–6
    thunder, 240–41
    volcanoes, 225–7, 230–33
    water, 244–54
    winds, 242–4
Ghost in the MP3, The, project 198
Gibson, William, 205
Gimzewski, James K., 76–8
Glennie, Evelyn, 16, 116, 301
Goldsmith, Mike, 163
Goodall, Jane, 142
Google, 197, 209, 211
Göpfert, Martin, 90
gorillas, 122
graphene, 79
grasshoppers, 90, 93
Great Animal Orchestra, The (Krause), 140, 142, 146, 149
Greenland Pump, 249
Guairá Falls, Paraguay/Brazil, 250
Guéguen, Nicolas, 41
Gurnett, Don, 273, 292–3

Hamby, William, 58
Handbook of the Senses: Audition, 109
Harbisson, Neil, 208–9, 211
Harmonices Mundi Libri V (Kepler), 296
harmonics, 65–9, 67
Harmonikon (Ptolemy), 295

Hart, Mickey, 273
Harvard–Smithsonian Center for Astrophysics (CfA), 279
HAT-P-11b exoplanet, 276
Havana syndrome, 172–3
HD 189733b exoplanet, 291
headphones, 31–2, 35, 206–7
health, 49–51
    birdsong and, 108
    calming sound and, 48
    cellular health, 80–81
    excessive noise and, 29–31, 160–61, 164–6, 182–3
    hearing damage, 29–34
    mental health, 313–14
    sonic healing, 49–51, 81
    vibrational healing, 81
Heap, Imogen, 220
hearing, 18–26
    brain and, 19–20
    ears, 16–19, 25, 29, 127
    hearing aids, 33, 209
    hearing color, 208–9
    hearing damage, 29–34
    hyperacusis, 34
    misophonia, 34
    neuronal, 19–20
Hearing Solar Winds (album), 53–4
Heatherwick, Thomas, 221
Heffner, Henry E., 109
Heffner, Rickye S., 109
Helmholtz resonance, 176
Hemingway, Ernest, 53
Hempton, Gordon, 305–7
Hesse, Hermann, 10, 189
Heydon, Christopher, 296
high frequencies, 19–20, 30, 62–3, 109–10
Hitler, Adolf, 185
Højerup church, Stevns Klint, Denmark, 53
Honshu earthquake, Japan, 234
Horowitz, Seth, 73, 111, 149

How the Mind Works (Pinker), 189
How to Speak So that People Want to Listen (TED talk), 186
howler monkeys, 122
Hubble space telescope, 271, 276
human voice, 184–6
    deep voices, 124
    orators, 184–5, 302
Hunga Tonga–Hunga Haʻapai volcano, Tonga, 231
Hykes, David, 53–4, 65
hyperacusis, 34
HyperSonic Sound, 171–3, 215
Hypogeum of Ḥal Saflieni, Malta, 177
HYRISS (hyperreal immersive sound space) system, 214

Iceland, 163
In Pursuit of Silence (book, Prochnik), 303
In Pursuit of Silence (film, Shen), 152,
India, 165–6
Industrial Revolution, 152, 163
infrasound, 62, 97–8, 115–17
insects, 85–94
    acoustic defense, 89
    "aggressive mimicry," 88
    hearing insects, 90–91
    Johnston's organ, 90, 92
    predators, 88–9
INVISIO, 170
Islamabad, Pakistan, 167
Iyo One (audio computer), 211

Jacob, Céline, 41
James Webb Space Telescope, 262, 271
Japan, 223, 312–13
Jazz Singer, The (film), 200
Jodrell Bank, 278
Jupiter, 286, 288–91, 296–8
Jurassic Park (film), 201

katydids, 88
Kauai, Hawaii, 246
Keating, Dave, 175–6
kenjutsu masters, Japan, 243–4
Kennedy, John F., 185
Kents Cavern, Devon, 5, 312
Kepler, Johannes, 294–8
Khait, Itzhak, 83
Kim, David, 41
King, Martin Luther, 185
King, Stephanie, 135
Kogut, Alan, 258, 262
Kosko, Bart, 163–4
Krakatoa volcano, Indonesia, 225–7, 230–31
Krause, Bernie, 94, 114, 140–42, 145–6, 148–9
Krisp (AI-based sound utility), 210
Kronos Quartet, 273
Kroodsma, Donald, 104
Kubrik, Stanley, 201
Kwando river, Namibia, 73

L-Acoustics, 214
Lake Maracaibo, Venezuela, 241
Lake Vostok, Antarctica, 288
Lalitpur, Nepal, 167
Lambda-Cold Dark Matter model, 262–3
language, 5–6, 17, 32, 189–90
lar gibbon, 122–3
Large Igneous Provinces (LIPs), 231–3
Last Wave, The (film), 202
Le Guellec, Hélène, 41
Leesman Index, 221
Levitin, Daniel, 189
Life of Birds (TV series), 105
lightning, 241
Lincoln Meadow, Sierra County, California, 145
lions, 119
Listen to the Deep-Ocean Environment (acoustic monitoring project), 173

listening, 10–12, 32, 143, 308–9
  cocktail party effect, 32
  corporate listening, 12
  silence and, 308–9
Listening for Ancient Gods (Eneix), 179
Lituya Bay, Alaska, 253
lizards, 117
logging, 145–6
Lomax, Alan, 191
London, England, 70, 167
London SoundLab, 184
Long Range Acoustic Device (LRAD), 171–3
Lovelock, James, 295
low frequencies, 19, 62–4, 109
Lynyrd Skynyrd, 30
lyrebirds, 105–6

Macaulay Library, Cornell, 94, 95
Macnamara, Jim, 11–12
Maguire, Ryan, 198
Mangini, Mark, 119
Marco Polo, 235
Marriott, Steve, 219
Mars, 147, 285–6, 295–7
masking, 46–7
Massive Attack, 271
mathematics, 66–8
Mayer, Alfred M., 197
MBA mnemonic, 38–9
McAdams, Rachel, 159
McGurk illusion, 43
meditation, 308, 311, 313
Melodio, 219
Mercury, 282, 296–7
Merrillville, Indiana, 241
meteor impacts, 232, 236–9
Meyer, John, 213–14
Mickey Hart's Rhythmic Universe (film), 273
microbaroms, 98
Microsoft, 3, 207
Middlesex University, 106

midges, 92
Millennium Footbridge, London, 40
Milliman, Ronald, 40
Milton, John, 151
misophonia, 34
Mithen, Steven, 17, 189–90
modulating electrical fields, 92–3
Montealegre-Zapata, Fernando, 87–8
Moodsonic, 45, 47, 221
Moog, Robert, 140
Mooney, Aran, 136
Moradabad, India, 167
mosquitoes, 90
moths, 89–91
Mount Washington, New Hampshire, 243
Moving Pictures Expert Group (MPEG), 197
MP3 files, 197–9
Muir, Tom, 187
Mumbai, India, 165–6
Münzel, Thomas, 161
music, 10, 16, 22–3, 28, 35, 51–2, 68–9, 189–95, 215–18
  AI and, 218–19
  audio revolution and, 215–18
  background music (BGM), 40–42, 199–200
  computer-generated, 218–19
  ethnic, 68–9, 190–91
  film, 200–201
  folk, 191
  harmonics, 65–9, *67*
  intention, 194–5
  masking and, 47
  musical instruments, 193, *193*
  musicians, 24, 30, 51, 219–20
  productivity and, 51–2
  singing, 53–4, 192–3
  sound reproduction, 195–200
  therapy, 49–51
  as a weapon, 190
Music from the Galaxies (Terenzi), 271

Musicophilia (Sacks), 24, 34
Myerson, Jeremy, 221–2
MyNoise website, 241
Mysterium Tremendum (Hart), 273
mysticetes, 134

NASA, 239, 257–8, 272, 279, 285
    Radio JOVE, 273
Natural History Museum, London, 93, 145
Neptune, 286, 289–91
New Horizons probe, 291
New York, U.S., 163, 167
Newgrange, Ireland, 177
Newton, Isaac, 294–5
Nghia, Vo Trong, 221
Niagara Falls, Canada/United States, 249–50
nightingales, 96, 106, 139
noise, 152–67
    aircraft, 160–61, 163
    antisocial behavior and, 169–70
    benefits of, 163–4
    construction, 153–4
    health and, 29–31, 160–61, 164–6
    industry and, 154–5
    noise mapping, 165–6
    oceans, 155–9, 162, 173–4
    pink noise, 164
    rockets, 159–60
    technology and, 173–4
    transport and, 152–3, 165, 169
    urban, 8, 107, 153, 162–3, 165–9, *168*, 310
    war and conflict, 170–73
    as a weapon, 171–3
    white noise, 164
Noise (Kosko), 163
Noise Abatement Act, UK (1960), 166
noise-induced hearing loss (NIHL), 29–31
Norris, Woody, 171, 215
North, Adrian, 41

Nouwen, Henri J. M., 301
nuclear weapons, 170–71
Nyquist–Shannon sampling theorem, 196*n*

oceans, 126, 130, 155–8, 245
    ocean gravity waves, 245
Ocho Rios, Jamaica, 89
odontocetes, 133–4
offshore wind turbines, 157
oil and gas exploration, 156–7
oilbirds, 102
Oldman, Tim, 221
Olivier, Laurence, 124
Olympic National Park, Washington State, 305–6
On Being podcast, 305
1 Giant Leap (Catto & Bridgeman), 191–2
One Square Inch of Silence: One Man's Quest to Preserve Quiet (Hempton), 305
orcas, 99, 129, 135–6, 141
Organisational Listening Project, 8, 11–12
Orkney, Scotland, 175–6, 191, 243, 251
oscillation 55
Our Masters' Voices (Atkinson), 124
owls, 102

Parncutt, Richard, 17
parrots, 105
passive listening systems, 174
Payne, Roger, 137–8
peafowl, 117
Pelling, Andrew, 77–8
Peng, Zhigang, 235
Perseus cluster, 12–13, 279
Perseverance rover, 285
pet sounds, 118–21
Pinker, Steven, 189
Pipedown campaign, 199
pitch-shifting, 263

Planck telescope, 262–3
Planets, The (Holst), 272
plants, 81–5, 144–5, 221–2
  infestations, 144–5
Plato, 49
Pluto, 291
podcasts, 7
polar bears, 123, 128–9
Pollen, Francis, 312
Pompeii, F. Joseph, 215
porpoises, 129, 133, 157
praying mantis, 90
Prochnik, George, 303–4, 313
prokaryotes, 76
psychoacoustics, 196–7, 207
Ptolemy, 295
public address systems, 212–13
Puerto Vallarta, Mexico, 167
puffins, 98
Pythagoras, 6, 66–8, 294, 296, 298

Quiet (Cain), 8, 307
Quiet Parks International (charity), 306–7

rainforests, 74–5, 103, 144
Rajshahi, Bangladesh, 167
Ratcliffe, Eleanor, 108
Reclaiming Conversation (Turkle), 7
religion/spirituality, 294–5, 308
reptiles, 117–18
Research Infrastructures for Citizens in Europe (REINFORCE), 273
resonance, 63, 177–8, 193
  Helmholtz resonance, 176
  stochastic resonance, 164
reverberation time (RT), 48–9
Reznikoff, Iegor, 9, 193
rhinoceros, 116
rhythm, 22, 87, 98, 302
Riley, Terry, 273, 293
Ring of Brodgar, Orkney, 175
Rinri Therapeutics (UK), 33–4
roaring, 119–20
rodents, 113
Rome, Italy, 152
Ronson, Mark, 216–17, 219
Rooney, Wayne, 36
Roopkund, India, 246
Rothenberg, David, 95–6, 98–9, 103, 106–7, 139, 149
Royal Albert Hall, London, 180–81
Rugolo, Jason, 211
Russia, 170, 172
Russo, Matt, 272–3, 297
Rydstrom, Gary, 201

Sabine, Wallace Clement, 180
Sacks, Oliver, 24, 34
Saturn, 283, 286–8, 290–91, 296–7
SaxaVord Spaceport, Shetland Islands, 160
Scapa Flow, 251
Schaeffer, Pierre, 73
Schafer, R. Murray, 69, 162, 196
"schizophonia," 196
Schnupp, Jan, 19
Schröder, Robert, 272
scorpions, 91
Scotland, 160, 190–91
sea lions, 129, 133
sea noise *see* oceans
sea otters, 129
seals, 129–30, 133, 157
seismicsoundlab.org, 235
Sense of Silence Foundation, 173
senses, 42–4, 74
Shaanxi earthquake, China, 234
Shakespeare, William, 257
Shen, Patrick, 152
Shinrin-yoku (forest bathing), 312–13
shrimp, 132
Shure, 214
silence, 301–15
  listening and, 308–9
  singing, 53–4, 192–3

Singing Comet, A (ESA), 271
Singing Neanderthals, The (Mithen), 189
Sirbuly, Donald, 79
sirenians, 129
sleep deprivation, 165
smart speakers, 209–10
Smoot, George, 273
snakes, 117
Snook, Richard, 78
solar system, 276, 282–91, 296–8
"Song of Myself" (Whitman), 1
Songs of Insects website, 86
Sonic Experience: A Guide to Everyday Sounds (Augoyard), 69
sonic healing, 49–51
Sonic Sea (documentary), 159
Sonic Wonderland (Cox), 178
"sonification," 271–3
Sonnenschein, David, 202–3
sonocytology, 76–8, 81
Sony 360 Reality Audio, 206–7
sound, 55–61, 266–7
   association and, 36
   behavioral effects, 40–42
   cognitive effects, 37–9
   crossmodal effects, 42–5
   definition of, 16
   emotional responses to, 20–21
   harmonics, 65–9, 67
   masking, 46–7
   mathematics and, 66–8
   physiological effects, 28–35
   psychological effects, 35–7
   reproduction, 195–200, 205–6
   sound libraries, 94, 149
   sound pressure, 57–61
   in space, 13, 257–99
   speed of, 56–61
   subconscious and, 20–21, 23
   survival instinct and, 5, 28
   timbre, 64–9
   underwater, 126–7, 156
   in the womb, 4, 17–18

Sound Design: The Expressive Power of Music, Voice and Sound Effects in Cinema (Sonnenschein), 202
SoundCloud platform, 271
Sounds of Life, The (Bakker), 101, 149
Soundscape: Our Sonic Environment and the Tuning of the World, The (Schafer), 69
soundscapes, 42, 45, 69–71, 143–6
Soundtracker (documentary film), 305
soundwaves, 55–64, 69
Soviet Venera probes, 284
Sowden, Mike, 188
Space Audio website, 293
Space Project, The, 272
space rockets, 159–60
Spaces Speak, Are You Listening? (Blesser), 181
Spangenberg, E. R., 41
Spartacus (film), 201
spatial sound, 196, 203, 206–7, 212–15
speech recognition, 209–10
speed of sound, 56–61
Spence, Charles, 42–4
spiders, 91
Spong, Paul, 159
squid, 136–7
Star Wars (film series), 200
starlings, 99
Steinbrugge, Karl V., 234
stem cell therapy, 33–4
Stewart, Martyn, 94, 162
Sting, 159
Stipe, Michael, 191
stochastic resonance, 164
Stonehenge, 178
Storey, Owen, 292
storytelling, 187–8
Sumatran Rhino Rescue, 116
Sun Rings (Riley), 273, 293
Suno, 219
superadditivity, 43–4
superior olivary complex (SOC), 25

supervolcanoes, 231–2
surface transducers, 209
SYSTEM Sounds website, 271–2

tachinid flies, 88
Tactical Communications and
    Protective System, U.S. Army, 170
Tales of Travel (Curzon), 236
Tambora volcano, Indonesia, 230–31
Tchernichovski, Ofer, 100
tectonic plates, 229, 233, 251–2
temporary threshold shift, 30
Terenzi, Fiorella, 271
Tesla, Nikola, 15
Thatcher, Margaret, 124
Third Ear, The (Berendt), 10
This Is Your Brain on Music (Levitin), 189
Tidal (streaming service), 206
Till, Rupert, 178
timbre, 64–9
Times, The, 180
tinnitus, 29–30
Tippett, Krista, 305
Titan (Saturn's moon), 286–7
Toba eruption, 231–2
Tohoku-Oki earthquake, Japan, 235
Tokai University, 80
Tokyo, Japan, 163
Tolkien, J. R. R., 70
Tolle, Eckhart, 27
"Tom's Diner" (Suzanne Vega), 197, 198
Toop, David, 95
Tsar Bomba nuclear test, 170–71, 237
Tunguska event, Siberia, 238–9
Turkle, Sherry, 7

ultralow-frequency (ULF), 266
ultrasonic waves, 171
ultrasound, 62, 109, 113
UN Environment Programme, 167
UN High Seas Treaty (2023), 158
Universal Sense, The (Horowitz), 73, 149

universe, creation of, 258–61
Uranus, 286, 289, 291
urban noise, 107, 153, 162–3, 165–8, *168*

Vajont river, Italy, 253–4
Valdivia earthquake, Chile, 233
Vega, Suzanne, 95, 197
Vegavis iaai, 94
Venus, 283–4, 296–7
vibration, 54–5, 73–4
virtual reality (VR), 203, 207
"vocal learning," 100–101
voice synthesis, 209–10
voice-user interface (VUI) 210–11
volcanoes, 225–7, 230–33
Voltaire, 295
"von Kármán vortex street," 243–4
Vonnegut, Kurt, 192
Voyager space probes, 138, 272, 289, 292–3

water, 244–54
    fountains, 222–3, 246
    hail, 246–7
    melting ice, 248–9
    oceans, 126, 130, 155–8, 245
    rain, 246
    running water, 222–3, 247
    snow, 247
    tidal bores, 252
    tsunamis, 36, 108, 226–7, 233, 237, 252–4
    waterfalls, 249–51, *250*
    waves, 249, 251–2
Watson, Aaron, 175–6
Watson, Chris, 89, 94, 133
Way of Silence, San Giulio, Italy, 311
Weir, Peter, 202
WELL Building Standard, 183–4
Whale Anti Collision System, 173
Whale Music: Thousand Mile Songs in a Sea of Sound (Rothenberg), 139

whales, 99, 126–7, 129, 133–9, 156–7, 162
　whalesong, 96, 101, 137–9
What About Me? (film), 191
whipbirds, 103
whistled languages, 120
"whistlers," 292
Whitman, Walt, 1
Whittle, Mark, 264–5, 280–81
Why Architects Need to Use Their Ears
　(TED talk), 181–2
Why Birds Sing (Rothenberg), 149
Widder, Edith, 136
Wilson, E. O., 85
Woods Hole Oceanographic Institution,
　Massachusetts, 174

World Health Organization (WHO),
　31, 165
World Report on Hearing, 33
Worth Abbey church, West Sussex, 312
Wragg, David, 93
wrens, 104
writing, development of, 6

Yalch, R. F., 41
Yellowstone volcano, 232–3
YouTube, 197

Zanclean megaflood, 251
Zander, Benjamin, 194
Zen gardens, 223